THE NEW MIDDLE AGES

BONNIE WHEELER, *Series Editor*

The New Middle Ages is a series dedicated to transdisciplinary studies of medieval cultures, with particular emphasis on recuperating women's history and on feminist and gender analyses. This peer-reviewed series includes both scholarly monographs and essay collections.

PUBLISHED BY PALGRAVE:

Women in the Medieval Islamic World: Power, Patronage, and Piety
edited by Gavin R.G. Hambly

The Ethics of Nature in the Middle Ages: On Boccaccio's Poetaphysics
by Gregory B. Stone

Presence and Presentation: Women in the Chinese Literati Tradition
by Sherry J. Mou

The Lost Love Letters of Heloise and Abelard: Perceptions of Dialogue in Twelfth-Century France
by Constant J. Mews

Understanding Scholastic Thought with Foucault
by Philipp W. Rosemann

For Her Good Estate: The Life of Elizabeth de Burgh
by Frances A. Underhill

Constructions of Widowhood and Virginity in the Middle Ages
edited by Cindy L. Carlson and Angela Jane Weisl

Motherhood and Mothering in Anglo-Saxon England
by Mary Dockray-Miller

Listening to Heloise: The Voice of a Twelfth-Century Woman
edited by Bonnie Wheeler

The Postcolonial Middle Ages
edited by Jeffrey Jerome Cohen

Chaucer's Pardoner and Gender Theory: Bodies of Discourse
by Robert S. Sturges

Crossing the Bridge: Comparative Essays on Medieval European and Heian Japanese Women Writers
edited by Barbara Stevenson and Cynthia Ho

Engaging Words: The Culture of Reading in the Later Middle Ages
by Laurel Amtower

Robes and Honor: The Medieval World of Investiture
edited by Stewart Gordon

Representing Rape in Medieval and Early Modern Literature
edited by Elizabeth Robertson and Christine M. Rose

Same Sex Love and Desire Among Women in the Middle Ages
edited by Francesca Canadé Sautman and Pamela Sheingorn

Sight and Embodiment in the Middle Ages: Ocular Desires
by Suzannah Biernoff

Listen, Daughter: The Speculum Virginum and the Formation of Religious Women in the Middle Ages
edited by Constant J. Mews

Science, the Singular, and the Question of Theology
by Richard A. Lee, Jr.

Gender in Debate from the Early Middle Ages to the Renaissance
edited by Thelma S. Fenster and Clare A. Lees

Malory's Morte Darthur: *Remaking Arthurian Tradition*
by Catherine Batt

The Vernacular Spirit: Essays on Medieval Religious Literature
edited by Renate Blumenfeld-Kosinski, Duncan Robertson, and Nancy Warren

Popular Piety and Art in the Late Middle Ages: Image Worship and Idolatry in England 1350–500
by Kathleen Kamerick

Absent Narratives, Manuscript Textuality, and Literary Structure in Late Medieval England
by Elizabeth Scala

Creating Community with Food and Drink in Merovingian Gaul
by Bonnie Effros

Representations of Early Byzantine Empresses: Image and Empire
by Anne McClanan

Encountering Medieval Textiles and Dress: Objects, Texts, Images
edited by Désirée G. Koslin and Janet Snyder

Eleanor of Aquitaine: Lord and Lady
edited by Bonnie Wheeler and John Carmi Parsons

Isabel La Católica, Queen of Castile: Critical Essays
edited by David A. Boruchoff

Homoeroticism and Chivalry: Discourses of Male Same-Sex Desire in the Fourteenth Century
by Richard Zeikowitz

Portraits of Medieval Women: Family, Marriage, and Politics in England 1225–1350
by Linda E. Mitchell

Eloquent Virgins: From Thecla to Joan of Arc
by Maud Burnett McInerney

The Persistence of Medievalism: Narrative Adventures in Contemporary Culture
by Angela Jane Weisl

Capetian Women
edited by Kathleen Nolan

Joan of Arc and Spirituality
edited by Ann W. Astell and Bonnie Wheeler

The Texture of Society: Medieval Women in the Southern Low Countries
edited by Ellen E. Kittell and Mary A. Suydam

Charlemagne's Mustache: And Other Cultural Clusters of a Dark Age
by Paul Edward Dutton

Troubled Vision: Gender, Sexuality, and Sight in Medieval Text and Image
edited by Emma Campbell and Robert Mills

Queering Medieval Genres
by Tison Pugh

Sacred Place in Early Medieval Neoplatonism
by L. Michael Harrington

The Middle Ages at Work
edited by Kellie Robertson and Michael Uebel

Chaucer's Jobs
by David R. Carlson

SACRED PLACE IN EARLY MEDIEVAL NEOPLATONISM

L. Michael Harrington

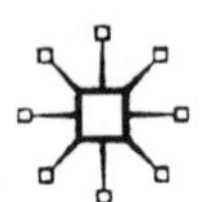

First published 2004 by
PALGRAVE MACMILLAN™
175 Fifth Avenue, New York, N.Y. 10010 and
Houndmills, Basingstoke, Hampshire, England RG21 6XS
Companies and representatives throughout the world

PALGRAVE MACMILLAN is the global academic imprint of the Palgrave Macmillan division of St. Martin's Press, LLC and of Palgrave Macmillan Ltd. Macmillan® is a registered trademark in the United States, United Kingdom and other countries. Palgrave is a registered trademark in the European Union and other countries.

ISBN 1–4039–6601–X hardback

Library of Congress Cataloging-in-Publication Data
Harrington, L. Michael.
Sacred place in medieval Neoplatonism / L. Michael Harrington.
p. cm. — (The New Middle Ages)
Includes bibliographical references and index.
ISBN 1–4039–6601–X (hc : alk. paper)
1. Place (Philosophy) 2. Sacred space. 3. Neoplatonism. 4. Philosophy, Medieval. I. Title. II. New Middle Ages (Palgrave Macmillan (Firm))

B105.P53H37 2004
203'.5—dc22 2004044537

A catalogue record for this book is available from the British Library.

Design by Newgen Imaging Systems (P) Ltd., Chennai, India.

First edition: October 2004
10 9 8 7 6 5 4 3 2 1

Printed in the United States of America.

For Kelly

CONTENTS

INTRODUCTION: NEOPLATONISM INSIDE AND OUTSIDE HISTORY

The visitor to Houston's Byzantine Fresco Chapel Museum sees something rather different from the medieval worshipper. The frescos housed in the museum originally occupied the dome and apse of a small chapel outside the village of Lysi, Cyprus.[1] Their new home in the Byzantine Fresco Chapel Museum is described by its promotional literature as a "reliquary box," but unlike a reliquary box, the museum is not designed to be seen from the outside. The parking lot leaves the motorist beneath a stone wall that shields most of the building from view. Her eye is drawn not to the building, but to the surrounding wall and a pool of water outside the main doors. The pedestrian approaching on foot from the nearby Rothko chapel is prevented at first by a line of trees from seeing the museum at all. The path to the main doors takes her off the street and under the shadow of the stone wall that surrounds the building, then to the main doors and the pool of water. An unobstructed view of the building can only be gained by walking into the residential neighborhood across the street. From this unlikely vantage, the museum reveals itself to be a large, apparently windowless cube, paneled with gray rectangles. The comparison with a reliquary box is instructive, for while a reliquary box, like most products of human art, makes use of geometrical forms, it does not strive primarily to incarnate them. Frequently jewel-encrusted or covered with sacred figures, it strives to be a *reliquary* box, and so to impress the viewer with indications of its sacred office. The museum takes the more distinctively modern approach of eliminating characteristics that we traditionally associate with buildings—the facade, for instance—in favor of a purer incarnation of geometrical form. Its effect, though, is muted by its retreat behind the line of trees and the surrounding stone wall.

Though the museum refrains from participating in the exterior space of the neighborhood, it carefully organizes its interior. The visitor enters a black steel box in the shape of an upside down shoebox, suspended around two feet from the concrete walls of the surrounding building. A skylight in the roof of the building allows daylight to wash down the concrete wall outside the box, "dematerializing the concrete wall," as the promotional literature describes it. Within the black box is the building's focus, a Byzantine chapel made of opaque glass panels that do not quite touch each other. They do not support the frescos housed within them, and they do not serve the more basic structural purpose of holding each other up. Instead, black shafts attach them for support to the walls and roof of the box. The chapel and its frescos are lit with florescent bulbs hidden beneath the floor, and incandescent bulbs hidden higher up in the corners of the chapel itself. Because there is no obvious light source, the walls and the frescos seem to glow of their own accord.[2] The overall character of the glass walls and the lighting is what the museum's literature describes as "immaterial materiality." The walls are visible but they hold nothing up. The chapel is pervaded with light, but it has no visible source. The interior provides the visitor with the illusion of a world made entirely from light itself.

"Immaterial materiality": this is the language of the Neoplatonists, used by them to describe entities that can be shaped as matter is, but without possessing properly material characteristics. The Neoplatonists typically reserve the phrase for truly immaterial entities like the soul, which is shaped as though it were matter by the thoughts it thinks. Yet the phrase is easily applied to some of the visible components of the world we see. The slender, luminescent walls of the museum chapel are immaterial in comparison with the thick, masonry walls of the Lysi chapel. The designers of the museum chapel have used this seeming immateriality to highlight the absence of the original chapel, but they have employed a method that is distinctively modern: the concealment of the bodily character of the building through the use of synthetic materials and new technologies. Though the historical school of Neoplatonism had neither the technology nor the inclination to employ this method, its name has become attached to it, and criticisms of the method have at times also become criticisms of Neoplatonism. This association leaves an important question unanswered: if the medieval builders of the Lysi chapel were to possess the technology needed to build the fresco museum chapel, would they choose to do so?

Neoplatonism Outside History

We find a demonized Neoplatonism inhabiting the margins of Vincent Scully's 1991 history of Western architecture, *Architecture: the Natural and the Manmade*. Scully begins this work by revisiting the themes of his earlier work on ancient Greek architecture, *The Earth, the Temple, and the Gods* (1962), in which he identifies a largely prehistoric form of architecture in the Greek temple. The temple architecture of Scully's archaic and classical

Greece takes careful account of the landscape in which it is placed. This landscape is not scenic, but sacred, "not a picture but a true force which physically embodied the powers that ruled the world."[3] The temple has only to gather and shape the already sacred characteristics of the landscape in their relation to the human worshippers and to the gods embodied in the temple. Each temple has characteristics specific to the god it embodies as well as its specific relation to the local landscape. Some temples do enclose an interior space appropriate to their god—the chthonian labyrinths beneath some temples of Apollo, for instance—but they do not do so to the extent of ignoring the landscape outside. Scully's Greeks neither ignore nor control their landscape, but contrive an architecture that can nestle within it.

Contrast Scully's Greeks with the Romans who follow them. The Roman temples that open out onto a landscape no longer incorporate themselves within that landscape, but attempt to dominate it.[4] While Greek temples are rarely situated on the top of mountains, nestling instead in a subordinate position, the Roman temple sits atop the mountain with a commanding view of the surrounding terrain. The temple does not constitute a focal point; it is designed as the source of sight, not its object. And so the space that it organizes is really the expanse of lands which can be seen from it. The stage is set for the marking off and appropriation of the lands viewed by these "commanders of the vista."[5] As it marks off the landscape to be controlled, the temple at the same time loses its power to be the embodying mystery of the god that appears in it. For the temple is now not seen as a place where anything appears. It is a control center.

Not every Roman temple opens out onto a space to be controlled, but even the temples that close in a space manifest their own transformation of nature. These temples close out and reject the exterior nature, and dematerialize the interior nature in a way that transcends the nature outside. Scully looks to the eastern Roman Empire for his primary example of this: the Church of the Holy Wisdom or Hagia Sophia in present day Istanbul. The Hagia Sophia is "a perfect embodiment of Neoplatonic wisdom in its condensation of the ideal shapes of circle and square,"[6] organizing the nave of the church around the square, and then topping the square with a circular dome. The goal of this geometrical organization seems to be the presentation of a vast space without body of any kind. The walls of the nave retreat behind colonnaded galleries. The dome floats above a series of windows, seeming to require no support at all. This "Platonic perfection of circle and square" convinces us "that it is more essential, more true, than the appearances of the natural world can possibly be."[7] Scully considers the Hagia Sophia to be no mere historical example of architecture, for "it was Hagia Sophia that set the course of European architecture for a thousand years." This architecture wrapped itself around the two principles we have described: "to control nature," by building so as to command vistas, "or to keep it out" by enclosing an interior space.

For Scully, the devolution of Western architecture begins with the appearance of philosophy, which he treats as functionally identical with the appearance of Platonism. The Greek approach to architecture, so amply described in *The Earth, the Temple, and the Gods*, collapses "almost overnight" with the appearance of philosophy in the middle of the fifth century before Christ.[8] Socrates is born in 470, Plato in 427. The Greeks continue to build temples, but their structure is now "stiff, linear, and dried out." The Platonists in particular are well paired with this geometrization of the Greek temple because of "the abstraction of Platonic thinking," which tends to deprecate "the very physicality of the older temples." The Platonic philosophers who take up the number theory of the Pythagoreans describe the maker of the cosmos as following a geometrical model when constructing it. The application of this insight to the craft of sacred architecture leads to a homogeneity in temple design. The specificity of the temple to its god and to its landscape begins to disappear in favor of the presentation of a universal in visible form. The elevation of the universal necessarily downplays the material character of the temple, since the material of a thing renders it here and now, in this specific place at this specific time. The geometrical method does not ensconce the temple within its local landscape, but imposes on that landscape, as best it can, a being of a different order: the geometrical object. While a geometrical object can never properly become visible, since it is universal, and the visible is always specific in character, place, and time, the attempt to incarnate a geometrical object in visible form can degrade both its material character and the specificity of its surroundings so as to make them more amenable to human manipulation. The regular is easier to control. The term "Platonic," then, as Scully uses it here, does not describe a school of thought as much as a method of design.

The twentieth-century awareness of a gradual geometrization of human-building methods and general organization of the landscape is not limited to the study of architecture. The father of the phenomenological movement in philosophy, Edmund Husserl, commented in 1936 on "the surreptitious substitution of the mathematically substructed world of idealities for the only real world, the one that is actually given through perception."[9] His student, Martin Heidegger, joined him in lamenting the dominion of the mathematized world, and took the further step of rediscovering a world not structured on a mathematical model, but grounded in what Heidegger calls the "place" (*Platz*) or "locale" (*Ort*). Heidegger was joined by Mircea Eliade, who more thoroughly and influentially explored the sacred character of places for twentieth-century anthropologists. For these two seminal thinkers in the twentieth-century rediscovery of place, Western history is the story of the loss of place, a loss that begins when the world is approached with a method that Heidegger explicitly identifies as Platonic.

Martin Heidegger

In the winter term of 1942 and 1943, Martin Heidegger taught a course at the University of Freiburg entitled "Parmenides and Heraclitus." His choice of two of the earliest philosophers in the history of Western philosophy allowed him the chance to reflect on the relation of history and philosophy, and his course notes leave us a concise three-part philosophical history. Two-and-a-half millenia now separate us from Parmenides and Heraclitus, both of whom wrote during the sixth century before Christ, but "the passing of the years and the centuries has never affected what was thought in the thinking of these two thinkers."[10] Their thoughts remain the same not because they stand outside of time, but because "what is thought in this thinking is precisely the historical, the genuinely historical, preceding and thereby anticipating all successive history." The thought of Parmenides and Heraclitus occupies no historical moment in time. Their thought is historical precisely because it precedes history in our usual sense of the word. Heidegger gives this prehistoric period the title of "being and word."[11] "Word" here refers not to anything spoken, but to the general context of myth that constitutes the Greek world. This context takes shape in the stories the Greeks tell about particular places and the things that belong in those places. When a Greek tells a story about a certain place or a certain thing, he gives meaning to that place or thing, and he also gives meaning to his own life, since he understands himself in terms of these same stories. Reality, or being, is found for the Greek only in the context of these stories and the meanings they provide to the landscape around him. Although Parmenides and Heraclitus write about being without always using the form of a story, the scope of their thought remains within the bounds of this mythic context.

"Western metaphysics, the history of the essence of the truth of beings from Plato to Nietzsche, comes under the title 'being and *ratio*.' "[12] Western history begins with Plato, who differs from his predecessors by taking up *ratio*, or "reason." Reason thinks from being to being, and this motion from one object to another already indicates why history begins with it. The storyteller takes what is old and retells it. The telling will always be different, yet what it says is the same. Reason, on the other hand, seeks to move slowly through the comprehension of one thing after another. It seeks the new. Heidegger blames this reason for the transformation of things and places into quantities, a slow measuring out of the landscape as human beings seek to comprehend it, literally, piece by piece. The history of this transformation is the history of Platonism. Not the history of its schools in the Athens and Alexandria of Late Antiquity, in which Heidegger shows no particular interest, but the rational approach to the world whose result is the history of Western Europe. Heidegger's Platonism is transcendental and ahistorical, a method rather than a school.[13]

Heidegger concludes the history of Western metaphysics with Nietzsche, who died in 1900, while Heidegger's own work inaugurates not something new, but the renewal of the old. He cautions his audience not to think of his work as the latest in a historical sequence: "I am not saying anything new here, as no thinker at all may be the slave of the pleasure to say the new. . .Essential thinking must always say only the same, the old, the oldest, the beginning, and must say it primordially." This third and last division of history—Heidegger names it "being and time" after his 1927 work of the same name—revisits the beginning. It is, then, the end of history, its destiny. This destiny is only possible in the West, where the problem of being has always been posed, and is only possible for a certain historical people: Heidegger's own Germans, the "people of poets and thinkers." The Germans are capable of evading the measuring out of reason in favor of a more primordial way of being in the world, constituted by a different mode of thought. Now they must fit themselves "into the place of the destiny of the West, a destiny that conceals a world-destiny." This is 1942, a particularly fateful moment for both the West and the world. Germany has laid siege to Stalingrad, and is facing Allied offensives in North Africa. Heidegger cautions his audience that German victory lies not in force, but in "a more primordial experience." How will the Germans gain a more primordial experience? Not by returning to the history of the West. Heidegger himself intends not to take up his own history, but to return his historical people to their beginnings, to a mode of thought which, unlike Platonic reason, does not measure out and consume, but gathers, binding the people together in their place. In this context, it becomes easy to see why Heidegger's rectorial address of 1933 prompted this puzzled response from Karl Löwith: "it was not quite clear whether one should now study the pre-Socratic philosophers or join the SA brownshirts."[14] This blurring of the pre-Socratic philosophers with the *Sturmarbeiteilung*, the National Socialist militia that brought Adolf Hitler to power, underscores Heidegger's abandonment of history, in which the *Sturmarbeiteilung* and the pre-Socratics stand 2000 years apart, in favor of a continually refreshed mythic "beginning."

Mircea Eliade

If Heidegger rediscovers the concept of "place" for the twentieth century, Mircea Eliade does the same for "sacred place." His is a more original discovery, since "sacred place," unlike "place," has never before been an independent theme in Western thought. His chapters on sacred place from *Patterns in Comparative Religion* (1949) and *The Sacred and the Profane* (1957) open up an entire field of theoretical anthropology. Like Heidegger, Eliade sets for himself a paradoxical task. He is a historian of religion, and aims "to understand, and to make understandable to others, religious man's behavior and mental universe."[15] But the historian faces an ever more difficult task as Western history progresses, for the forms of religious life that survive depart more and more from the essence of "religious man." We cannot even look to the religions of the recent past—Christianity, Hinduism, Confucianism—for better examples, since they possess a "large written sacred literature." All this writing ensures that they are "too clearly marked by the long labor of scholars." Eliade implies that scholarship not only does not reflect the actual practice of the religion, but that it actually steers the course of the religion away from its essence. When the scholar writes about her religion, she takes a step back from the experience of that religion, and those who read her work become accustomed to thinking about their religion rather than experiencing it. Eventually, the experience is corrupted by the scholarship. The corruption of scholars has shaped Christianity almost from its first appearance, and so Christianity is not suited to give us the essence of "religious man." Eliade turns instead for inspiration to the rural peasants of Europe, whose religious practices "represent a more archaic state of culture than that documented in the mythology of classic Greece." Not that these peasants are not Christians, but they transcend the "historical forms of Christianity" in favor of a "primordial, ahistorical Christianity" that is cosmic in character. Like Heidegger, Eliade conceives the removal of history as essential to his task. A religion becomes historical when it begins to accumulate texts. These texts become the province of the literate, and the literate, by their very literacy, depart from the essence of "religious man." As a religion becomes historical, then, it undergoes a process of devolution.

We now stand at the end of that process, and suffer its explicitly spatial consequences. The scholarly reinterpretation of religion has made reason a fundamental part of religious practice. The interpretation and codification undertaken by the scholar already constitute a form of science, with its own methods and goals. Such a science stands a long way from our contemporary natural sciences, yet the reason at work in the two is the same. This is important if we wish to understand how historical religion devolves in a spatial way, for it is "scientific thought" that leads to the "gradual desacralization of the human dwelling."[16] A modern house is simply "a machine to live in," which for Eliade means that "it takes its place among the countless machines mass-produced in industrial societies."[17] Eliade's religious peasant, on the other hand, conceives the house not as a machine, but as the center of his world. It strives to embody in itself the characteristics of the local temple. For the religious man, both house and temple are sacred places, and so break up the otherwise homogeneous space in which human beings move. Their sacred character provides an orientation through which all other motion becomes comprehensible. For "profane experience, on the contrary, space is homogeneous and neutral; no break qualitatively differentiates the various parts of its mass."[18] In homogeneous space, objects have position only relative to each other, just as in geometrical space. One point of a triangle may be above the base, but none of its points is "above" without qualification. Their space has no center from which such a direction could be meaningful. In the absence of the orientation given by a sacred place, a human being moves only as "governed and driven by the obligations of an existence incorporated into an industrial society."[19] Industrial society is the power at play in homogeneous space, though it is not the creator of homogeneous space; it seems to move into the gap created when the world grounded by sacred places disappears. Homogeneous space is simply the consequence of "desacralization," the name Eliade gives to the method by which sacred places are erased. He wisely does not refer to this method as Platonic, and he does not attribute it to a particular historical school, since, as we have seen, history itself is part of the problem. He refers the concept of homogeneous space only to "the common stock of philosophical and scientific thought since antiquity."[20]

Neoplatonism Inside History

Not every recent work on "place" and "sacred place" has identified the geometrical model of space as the moving force within Western history, along with an ahistorical utopia as its remedy. Edward Casey's 1996 work *The Fate of Place*, for instance, charts carefully the interaction between "place" and "space" *within* the course of Western history.[21] Casey restricts his discussion to "place," and does not address those authors who only take up the topic of the sacred place. His approach is well adapted to the Neoplatonists, for whom the sacred place is not a subcategory of place, but requires an independent treatment. The two concepts work against each other, and it is only when one of them disappears, or the difference between them is elided, that the stage is set for the dematerialization of place that now seems so pernicious. There is, then, a gap in contemporary scholarship on "place" and "space": the Neoplatonic concept of sacred place has not been explicated or set in its proper historical context. The present work aims to fill that gap.

Caveats Concerning the Scope of the Present Work

Contemporary theoretical anthropology has had to come to terms with the difference between what we say about a place and how we act within that place. Our actions within sacred places may differ markedly from what we think and say about them. In what follows, I will only make reference in passing to observed practices within sacred places, which present grave problems for the scholarly interpreter and properly belong to the realm of ethnography. I will look instead at reflections on place, specifically as we find them in the texts of the Neoplatonists who take up the theme in Late Antiquity and the early Middle Ages, a period extending for our purposes from around A.D. 200–900. In focusing on philosophical texts, I abandon any attempt to chart the actual practice of a Christian lay member of a church. The figures we will look at—Plotinus, Iamblichus, Maximus, Eriugena, Cusa—were all men, none of them married, all of them well educated in the philosophical tradition. We know nothing about the central figure of this study, Dionysius the Areopagite, save that he probably lived in Syria around five hundred years after Christ, but he likely belongs to the same group. Were we interested in the practices of the early medieval period, such a rarefied group could give us only a narrow picture. As contributors to the tradition of reflection on the sacred place in the Middle Ages, however, they constitute a coherent and central group. The later figures in this study read many of the earlier figures, and, more importantly, made them central to their own reflections. Because their reflections directly influenced each other through the transmission of their texts, we learn a great deal about the philosophical development of the early medieval period by focusing on them without constant reference to historical practices of the time. This is not to say that the development of practices within the larger community has no effect on the philosopher's reflection, but the two interact only indirectly and in unpredictable ways.

This difference between the text of the philosopher and the act of the lay worshipper suggests a second caveat. The philosophers we will examine occasionally refer either to specific buildings, or to specific types of buildings. Iamblichus, for instance, discusses the specific structure of the oracles at Delphi, Didyma, and Claros. Dionysius discusses the generic structure of the Christian church. Churches which do not differ significantly from his description are still in active use today. Likewise, the ruins of the oracles at Delphi, Didyma, and Claros still exist. Can we derive the descriptions of Iamblichus and Dionysius from the buildings themselves? Or can we derive the meaning of the buildings from the descriptions given them by Iamblichus and Dionysius? In the work that follows, I will treat the buildings as though the answer to both these questions were "no." In doing so, I hope to avoid turning this into one of the "countless studies" that "continue to perpetuate the fiction that buildings do have inherent, stable meanings."[22] The builder of a sixth-century church in certain areas of the Mediterranean basin could have read Dionysius' work *On the Ecclesiastical Hierarchy*, and designed the church to provide for the activities mentioned by Dionysius in that text. But the same church plan could have been discovered by other means, and the plan does not govern the attitude and expectation of the people who perform the rites in the completed church. The relation between the building and the philosopher is as unpredictable as that between the philosopher and the surrounding community.

Our discussion is further complicated by the fact that authors and translators do not mean the same things by "place" and "space." In the ancient and medieval authors, I have always translated *τόπος* and *locus* as "place," and *χῶρος* and *spatium* as "space." I have done this even in cases where the author uses *τόπος*, for example, but means what we would describe as "space." The meaning, in these cases, should be clear from the context in which the term is used. While I have translated most ancient and medieval texts myself, I have relied on existing translations for modern and contemporary works, always checking them against the original texts and silently correcting them where necessary. Heidegger's translators have not always been careful to preserve his distinction between place (*Platz*) and space (*Raum*). Where they do not preserve it, I have corrected them. Eliade, at least in William Trask's translation, uses "place" and "space" indifferently in his discussion of the sacred. I have allowed his ambiguity to remain, but when I discuss him in my own words, I use the term "place" to describe his "sacred space," and I use "space" when he speaks of "homogeneous space," since in Heidegger's now traditional distinction only space can be homogeneous. In general, I have tried to alert the reader on every occasion where an author uses a place-related term in an unusual way.

The Plan of the Present Work

I will begin by surveying the reemergence of place as a topic for philosophical reflection in the twentieth century. We will see how a reading of Aristotle's *Physics* inspires Heidegger to rediscover the concept of "place" as distinguished from "space." While space has no effect on the thing that occupies it, the place of a thing determines whether and how it will reveal its nature. Heidegger's works written during the National Socialist period in Germany begin to stress language as the means by which things manifest their natures. Language causes all things to cohere in a "world," by which Heidegger means not just a totality of things, but their meaning for a historical people such as the Greeks or the Germans. I will prefer the broader term "context" when describing this network of meanings, since "context" has an explicit reference to language, but I do not mean to imply that the context is overlaid by a historical people on things that already exist. It may be that things as we know them cannot exist without a context. The reference to "historical people" quietly slips out of Heidegger's reflections on place after the demise of the National Socialist party. His lectures given in the 1950s identify place as the "gathering" of humans in their interactions with each other, but also with the earth, the sky, and the gods. Because the gods are gathered into every place, no sharp difference between place and sacred place can emerge. The sacredness of the place, though, is not imparted by the presence of the known gods of a historical people, gods that can be named, but unknown divinities who remain absent even when gathered in a place. The section will conclude with further reflections on the sacred place: the problem of naturally sacred places in Mircea Eliade and his critics, and the dependence of place on the human body, neglected by Heidegger but developed by Edward Casey and Henri Lefebvre.

The Neoplatonists are the first Western philosophers to include both place and sacred place in their reflections, and the changing relation they give to the two helps to explain the eclipse of place in the modern era, and to locate our contemporary rediscovery of place relative to the Western philosophical tradition. Plotinus and his fellow Neoplatonists strive to distinguish themselves from Aristotle in their understanding of place, but their own understanding of place reflects precisely what Heidegger sought to find in Aristotle, namely, that place "has a power," precisely because it is a "gathering." The fourth-century Syrian Iamblichus understands this as the gathering performed by a container. Place is not the passive shell around a body. It enters into the thing and gathers it together so that it does not disperse into nothingness like water emptied from a glass into the ground. The sixth-century Damascius also treats place as a gathering, but the container model is less interesting to him than the way a thing's place sets it in relation to a larger whole. In a human being, for instance, the heart may be said to have place if it is on the left within the context of the whole body. Take the heart out of the body and it may have a position, but not a place. A thing can exist outside of place, then, but it can fully become itself only when it is in its place.

While the Neoplatonists do not undertake a thematic discussion of the sacred place as they do for place in general, the sacred place becomes increasingly important in discussions of the soul's salvation. The Platonic identification of unity with goodness guarantees that, in some way, their philosophical systems will find our good as human beings in the achievement of unity with the gods, with ourselves, and with the cosmos. The later Neoplatonists famously deny that we can achieve such unity through the interior thoughts and prayers of our souls. We are embodied beings, and can only achieve unity with the gods within a place. Iamblichus explains how places become sacred in his major work, *On the Mysteries of the Egyptians, Chaldeans, and Assyrians*. Places become sacred within a mythic context, which superimposes a structure of interrelated symbols on an already existing cosmos of things and places. The laurel tree is a thing simply by having the form of a laurel tree, but it is sacred because of the mythic association with Apollo provided by the Hellenic religion. Iamblichus is careful to maintain a distinction between thing and symbol, between place and the sacred.

It is a fellow Syrian, the mysterious Christian who writes under the name of Dionysius the Areopagite, who undertakes to write a treatise devoted to the rites of the Christian church.[23] I do not treat "Christian Neoplatonist" as an oxymoron here. The term "Christian" most directly refers to a religious practice and a credal affiliation, while the term "Neoplatonist" refers most directly to a school of thought. Christians rely on some form of thought to understand their creeds and practices, and the Christians we will be speaking of here rely on Neoplatonism as the preferred framework for their creeds. When I wish to speak specifically of non-Christian Neoplatonism, I will use the phrase "Hellenic Neoplatonism." I have borrowed this phrase from Hilary Armstrong, who uses it for the same purpose.[24] He notes that this phrase is appropriate because the Hellenic Neoplatonists use it to refer to themselves. More traditional possibilities such as "non-Christian" and "pagan" are less than ideal. The term "non-Christian" indicates only opposition to Christianity, and not affiliation with the traditional rites of the Greek people. Likewise, the term "pagan" assumes that the Greeks have no formal religious practice comparable to the Christian rites (the term originally designated someone who was not a soldier, hence, to the Christians, someone who was not a soldier of Christ).[25] As we will see, the Hellenic Neoplatonists have their own complex and well-established rite.

Dionysius is familiar with the Hellenic Neoplatonists, especially some version of Plotinus and Proclus, and his philosophical system depends greatly on the work of Iamblichus. Like Iamblichus, Dionysius works within an already established ritual tradition. The Christians, like the Hellenes, have a temple and an altar, but they have moved the altar indoors, and the walls of the church now enclose the space within which things can become symbolic. Where Iamblichus simply says that the sacred place is the medium for divine union, Dionysius begins to explain how this is possible. He requires more of the lay Christian than Iamblichus requires of the Hellene. The Christian must not simply perform the actions of the rites in the sacred place, but also contemplate the symbols employed in the rites. The divine does not enter into the bodily to remain at the level of the bodily, but to draw the contemplator to the intelligible. This treatment of the symbols as objects of vision, with the goal of elevating the viewer to the immaterial, raises in our own context the specter of the dematerialization and manipulation of the bodily. These problems do not arise in Dionysius because his symbols must be thoroughly material to be effective (thereby resisting dematerialization) and function only within the walls of the church (and so resist manipulation for the sake of controlling the landscape).

We will follow up our study of Dionysius with a look at two authors who read his work with great interest. Maximus the Confessor provides the first lengthy discussion of the sacred place in his *Mystagogy*, a work inspired directly by Dionysius' *On the Ecclesiastical Hierarchy*. Dionysius, like Iamblichus, roots human nature firmly in the visible world. Maximus suggests in a number of works, including the *Mystagogy*, that human nature is prior to both the visible and the intelligible worlds. Human nature is a microcosm, containing every kind of reality within itself. Though human beings find themselves immersed in the visible and forgetful of the intelligible, they see in the church an image of the cosmos: both bind all things together without destroying the particular characteristics of each thing. By sharing in the activity of the church, human beings fulfill their microcosmic vocation. Because he foregrounds the likeness between the church and the cosmos, Maximus strains the distinction between "place" and "sacred place." His ninth-century Irish translator Eriugena strains the distinction even further, suggesting that, since we stand outside the distinction between the created and divine, we may see each created thing as divine. And we are right to do this, since the substance of each thing has the preeminent characteristic of the divine: it is unknowable. A brief look at Abbot Suger, whose renovations to the abbey church at Saint-Denis have long been considered as the origin of Gothic architecture, will conclude the section.

The later Middle Ages do not overlook the topic of place. It appears frequently, but as a reassertion of Aristotle, and not a direct engagement of the Neoplatonists. The major step from place to space in this period is taken in Étienne Tempier's famous condemnations of 1277, which threaten teachers at the University of Paris with excommunication if they teach that things have natural places. The suppression of the teaching of natural places paves the way for the abandonment of place altogether. Tempier phrases his condemnations in the language of the thirteenth century's revived Aristotelianism, and Edward Casey has ably discussed their effect in *The Fate of Place*.[26] A more Neoplatonic engagement with place and sacred place does not reappear in earnest until the fourteenth and fifteenth centuries, when the fate of place has been sealed, and the tension between place and sacred place is no longer strained but overcome. We will look briefly at Nicholas of Cusa here, an unparalleled figure for our purposes, since he combines an interest in the early medieval Neoplatonists with an equal interest in the emerging mathematization of things. Cusa continues to treat human beings as microcosms, but the world they embody is in every case their own world. I cannot experience the world, or perspective, of other historical peoples, or the world of other species, but I can believe that they have their own worlds, though they are hidden from me. Cusa's famous "learned ignorance" rests on the discovery that my world is not the only world, though my knowledge is limited to that world, and that the possibility of multiple worlds hangs on a being who can see from the perspective of them all. I experience this union of perspectives only negatively, as a form of ignorance. Cusa employs the concept of homogeneous space only to provide a field in which these many different perspectives can come about.

The Neoplatonic sacred place serves as the medium for our union with ourselves, other beings, and the divine. The need for that union does not disappear along with the medium that provides for it. We find in the absence of place the full development of what we now call "feeling" or "emotion." Among the ancient schools of thought, the manipulation of emotions is necessary to preserve the health of a human being, but in the absence of place, it takes on the more fundamental role of uniting us with each other, the cosmos, and the divine. I will not attempt to survey the development of this theory of the emotions—that job belongs to a history of romanticism—but I will track the influence of this theory on several scholars whose work figures in the main body of the text, among them Pierre Hadot and Lindsay Jones.

The reader who seeks only to explore the history of place and sacred place as concepts in medieval Neoplatonism may comfortably read the middle three chapters of this book without sensing an incompleteness to the argument, and without encountering the confrontation with the present day posed in the first and last chapters. The reader who, on the other hand, wishes actively to restore a sense of history to contemporary thinking about the sacred place will profit more from a reading of the book from beginning to end.[27] The juxtaposition of the present day with the medieval here is not arbitrary. The present day and the medieval stand on opposite sides of modernity, but they share a relation to it. The medieval holds the seeds of modernity, while the present day holds its fallen leaves. In both periods, we see a tension between place and the sacred, which is not to be found among the moderns. We may more effectively begin with the present day, because the exigencies of our own situation have made it necessary to articulate the tension that the early medieval Neoplatonists only hint at.

CHAPTER 1

THE REDISCOVERY OF PLACE

It is a sad fact that Martin Heidegger and Mircea Eliade, two thinkers who have revitalized the study of place in the twentieth century, both patronized regimes now remembered only for their brutality. Heidegger's involvement with the National Socialist movement in Germany is well known, and Eliade's support for the Romanian Iron Guard is gradually getting more attention.[1] The most unsettling aspect of their involvement with these regimes, for our purposes, is that it may possibly explain their interest in the concept of place. Especially in essays written and courses taught during the National Socialist period in Germany, Heidegger emphasizes place as grounding the world of a historical people. Eliade speaks repeatedly of the degradation of modern industrial society where, "properly speaking, there is no longer any world."[2] Industrial society produces houses and people without distinctiveness in a space which is likewise homogeneous and undistinguished. Did Eliade look to Corneliu Zelea Codreanu, the head of the Iron Guard, as someone who could forcefully reimpress a center on a world become homogeneous?

This world-grounding place is neither the beginning nor the end of the twentieth-century rediscovery of place, and relies on an older and more fundamental conception of place as what allows a thing to reveal itself in the fullness of its nature. While we may never encounter the thing outside the context of our own historical people—with the memories, habits, and myths that shape our experience of things—these encounters simply make culturally specific the formal structure of the "natural place." Later authors like Henri Lefebvre and Edward Casey will more fully develop this kind of place by drawing out its essential connection with the body, but it is Heidegger who reintroduces it to the philosophical tradition. Since Heidegger is not only the first to make place a major theme after the Middle Ages, but is also responsible for "the most suggestive and sustained treatment of place in this century,"[3] I will rely on him as the primary source of twentieth-century thinking about place.

The Heideggerian Place

During the winter term of 1924 and 1925, the young Heidegger taught a course on Plato's *Sophist* at the University of Marburg. His lectures for the course helped to shape the work Heidegger was writing at the time—*Being and Time*—which was to be published in 1927 and remains Heidegger's most influential work. The *Sophist* course lays out for the first time the character of place, which Heidegger was to develop in *Being and Time* and in later works throughout his life. Heidegger did not teach the course with the intent of writing a book as he went, but he retained his notes for the course, and several of his students took notes of his lectures. Later in life, Heidegger allowed his lecture courses to be published in book form, a project that continues today. The *Sophist* course appeared in 1992, some fifteen years after Heidegger's own death.

Place

Heidegger's discussion of place in the *Sophist* course takes its point of departure from Aristotle, who says in his *Physics* that place "has a certain power" (ἔχει τινὰ δύναμιν).[4] When a fire is lit, it rises upward, striving to attain the height of the stars above, while a clod of dirt falls downward until it hits the earth or some intervening body. The fire and the dirt do not initiate these motions. The natural place of each exerts a power over it and draws the fire or dirt to itself. Aristotle identifies the natural place of fire and dirt in the context of a finite universe with the earth at its center. The place of dirt, for instance, is the center of the universe: the absolute "below." Though Heidegger's twentieth-century students cannot think of the universe as geocentric, he cautions: "we may not permit mathematical-physical determinations to intrude" on Aristotle's theory. Heavy things still fall, and if we attribute this action to gravity rather than the power of place, we must nevertheless acknowledge that such things can fully reveal their nature only when they have reached their natural place. This is the power of place in Heidegger's *Sophist* course: not the power to cause motion, but the power to make a thing properly present. Place, he says, "constitutes precisely the possibility of the proper presence of the being in question." Place is able to make a fish, for example, present to us in the water, which is its proper place. If an angler catches the fish and leaves it to die on the bank, something essential to the fish is lost. Most of its possibilities for existence disappear, leaving only a certain shape and some material constituents: flesh and bone. Such a residue manifests very little of the fish's nature. By allowing a thing to become "properly present," place reveals the nature of the thing; this is its connection to the thing's possibilities and its very existence as such-and-such a thing.

Most modern scholarship has focused on a nearby passage in Aristotle's text, in which he defines place as "the limit of the surrounding body."[5] A fish swimming in the water, by this definition, has the surface of the water that surrounds it as its place. Heidegger does not ignore this passage in Aristotle, but neither does he regard it as Aristotle's definitive treatment of place. Heidegger introduces the "limit of the surrounding body" definition as a "further clarification" of the discussion of place, rather than a new or better theory. After describing its essentials, he claims that it is dependent on the natural place theory, saying, "this peculiar determination of place, as the limit of what encircles the body, is understandable only if one maintains that the world is oriented absolutely, that there are preeminent places as such."[6] Heidegger does not explicitly say why we will end up in confusion if we treat the two theories as separate, but we may conjecture that the "limit of the surrounding body" theory by itself does not justify our speaking about place as a category descriptive of a thing. The theory does not have the explanatory power of our ordinary use of place, as when we say that Socrates is in the marketplace. This statement gives us useful information about Socrates—it tells us where he is, and so it expresses what Aristotle calls the category of "whereness" (*ποῦ*)—while to say that Socrates is in a surrounding envelope of air tells us virtually nothing. If, on the other hand, we know that to be in the air and on the earth is the natural place of a human being, then we do know that Socrates is where he should be if he is to reveal the fullness of his nature. The elemental constitution of the surrounding body is now tied to the nature of the thing it surrounds, and so deserves examination as a category in its own right.

If the being of things has within it a determination to a specific place, then we do violence to the being of things if we consider them without reference to place. This violence, Heidegger suggests, motivates Aristotle's famous critique of Plato's separate forms. Plato thinks that he can treat "natural beings" (*φύσει ὄντα*) in the same way he treats geometrical objects, by extracting their principles from the realm of motion. But "in every category of physical beings there resides a determinate relation to motion." Leave out this relation, and you leave out an essential component of the being, with the result that the principle of the being, instead of acting from within the being itself, becomes something separate, an independently existing form. If we are to treat physical beings adequately, we must perceive along with the principle the characteristic of being moved, which then "must basically be something else as well, namely the *τόπος* itself whereby being and presence are determined."[7] If place is only the innermost limit of the surrounding body, this critique has no force. A fish on the riverbank, surrounded by air, has a place in this sense no more and no less than a fish in the water. Both have a body surrounding them, which cloaks them so as to leave no distance between the two. The theory of natural places is essential to the bonding of place and being in Heidegger's thought here. Only because of this bond can he repeatedly use "correct" and "proper" as modifiers of place. Heidegger may then conclude: "place is the ability a being has to be there, in such a way that, in being there, it is properly present."[8]

In the *Sophist* course, the nature of a thing includes a determination to a specific place. In *Being and Time*, the human being who puts the thing to use plays an essential role in giving the thing its place. Human beings manipulate and use things; they treat them as tools. To do this, they render them close at hand, and they orient them toward a specific goal. These two activities—putting nearby and orienting—give the tool its place. The character of the place, then, is not independent of the human being who puts the tool in its place. And human beings never organize only one tool at a time, so that "the actual place is defined as the place of this useful thing in terms of a totality of the interconnected places of the context of useful things at hand in the surrounding world."[9] Each thing has its place in relation to all the other things that go together to make up the world we live in from day to day. This world is not the Aristotelian cosmos, with its absolute "up" and "down." The proper place of a thing is now determined not by the natural organization of the cosmos, but by contexts supplied exclusively by human beings. This human-centered sense of place will be essential to Heidegger's later treatment of place as the ground of the world of a historical people: place gives meaning to things *for* people.

Position

Heidegger's treatment of place in his *Sophist* course arises in the course of an apparently unrelated discussion: the difference between geometrical and arithmetical objects. Specifically, he wonders about the difference between a monad and a point. Both are fundamental principles of their respective disciplines, the monad or "one" for arithmetic, and the point for geometry. They differ in only one respect: the point is a monad to which a "position" (θέσις) has been added. Heidegger knows that he is opening up a new field of questions by isolating "position" in this way, and in the *Sophist* course he begs off a thorough discussion of the term: "a thorough elucidation of this nexus would have to take up the question of locale (*Ort*) and space (*Raum*). Here I can only indicate what is necessary to make understandable the distinction of the ἀκριβές within the disciplines of mathematics."[10] For the purposes of his discussion here, Heidegger follows Aristotle in defining "position" as "orientation, situation; it has the character of being oriented toward something."[11] Position cannot exist by itself, but requires something else toward which it can orient itself. Its relational character provides an easy way for us to distinguish it from place, since place is not relational. The rock's place is the earth below us; it is not below to one person, but above for another person, depending on where they are standing. The rock may, however, have position in addition to place: it may be on one woman's right, but another woman's left, and so have a different position relative to the two women.

Beings, then, have both position and place. They are in their natural place, and have position relative to other beings. If we remove the power to make present from the place of a being, we are left not with a place, but a multiplicity of sites (*Lagen*). When this happens we no longer say that the being is in place, and we call it a geometrical object, not a being. Objects in sites retain their position. For example, if I remove the power to make present from the place of a tabletop, and am left with a square, a geometrical object sited in no particular place, I may still distinguish the "top" of the square from the "bottom" of the square. The sides have different positions relative to each other. How do we then distinguish a point from a monad, the original question that prompts Heidegger's discussion of place in the *Sophist* course? The point occupies a site, which means it has position relative to the other points that I can posit. The monad, on the other hand, occupies no site, and so cannot have relative position.

Region

Heidegger does not mention "region" (*Gegend*) at all in his *Sophist* course. The necessity for a third term, in addition to "place" and "position," in his theory seems to have revealed itself to him only in the writing of *Being and Time*, in which it first appears as a distinct theme. This is not to say that the meaning Heidegger later gives to the term "region" is not present in the *Sophist* course, but it remains indistinct from the meaning of place. To separate the region from the place in the Aristotelian terms of the *Sophist* course, we would have to take a natural place, such as "the earth below," and separate the 'below' from the 'earth'. 'Earth' is the place; 'below' is the region.

In *Being and Time*, Heidegger tells us that regions precede places. A thing has its place when it is oriented toward a certain job and brought close to me in order to do that job. Most jobs require several things collectively oriented toward them. The relation of these things to each other is accomplished by regions. If I am to successfully type an essay, I need the mouse and mousepad "on my right," the keyboard and monitor "in front," and the chair "below." The combination of regions and places allows me to experience surroundings: "this regional orientation of the multiplicity of places of what is at hand constitutes the aroundness, the being around us of beings encountered initially in the surrounding world."[12] The fact that I go about my day surrounded, as it were, does not mean that I pay attention to my surroundings as such, or that I pay attention to any given region as such. When I reach for the mousepad I do not pay attention to "my right." In his most extended treatment of the region, the "Conversation on a Country Path About Thinking" (1944–45), Heidegger claims that a single region lies behind all these restricted places and regions.[13] We ordinarily pay no attention at all to this region, but it constitutes the ground of all the places and things to which we attend in our day-to-day lives.

Space

The *Sophist* course treats "space" (*Raum*) as a strictly modern term, probably identical with the "pure dimensions" that cover over the richer fabric of natural places. If we misunderstand place and region as position and space, then we become unable to distinguish sharply the geometrical from the visible world. Heidegger expects his modern audience to be in the throes of such a confusion when he reports Aristotle's claim that geometrical objects are not in place. He comments that, "taken in terms of modern concepts, this has the ring of a paradox, especially since *τόπος* is still translated as 'space.' "[14] The modern translator, not understanding the difference between place and space, renders Aristotle's claim as "geometrical objects are not in space," which is indeed paradoxical. Geometrical objects cannot be conceived without space, since they have dimensionality: a triangle requires a two-dimensional surface if we are to conceive it. Place, on the other hand, as the preserve of visible things, is denied to the triangle and to all other geometrical objects.

In *Being and Time*, Heidegger finds a meaning of space that does not make it responsible for confusing the geometrical and the sensible. Our relation to space in Heidegger's sense determines whether or not we will be able to establish places and regions. When space is merely found, or "discovered non-circumspectly by just looking at it, the regions of the surrounding world get neutralized to pure dimensions."[15] This neutralization results in the confusion of geometrical with sensible. But space need not be merely found. It can be given, an act which Heidegger describes as "letting innerworldly beings be encountered in a way that is constitutive for being-in-the-world."[16] I give space by giving to entities within the world the possibility of being encountered, and through this encounter, I come to be in the world. Space, then, belongs to both me and my world. "*Space is neither in the subject, nor is the world in space.*" Space is not in the subject, meaning that space is not a filter in us by which we shape the data we receive through our senses. We are not isolated from the world in such a way that we need filters of this kind. As Heidegger has already explained, our "being in the world" is a special kind of "being in," one in which neither we nor the world could exist or be comprehensible without the other. And the world is not in space, meaning that space is not a neutral field, or "pure dimensionality," capable of containing a world of things. Pure dimensionality is characteristic of geometrical objects, not sensible ones, as we have seen.

Heidegger's conception of "giving space," first proposed in *Being and Time*, is the central theme of his 1951 essay, "Building Dwelling Thinking," in which he develops it into the more encompassing concept of "dwelling." Dwelling, like the giving of space for things, is not something that we recognize very often, and so it merits a little explanation. Dwelling is identical with building, but this building can take two forms: cultivation and construction. Cultivation only "tends the growth that ripens into fruit of its own accord," and it is distinguished from construction by this very fact: that it "is not making anything."[17] The vintner who tends the grapevines does not make the grapes, but, by "preserving and nurturing" the vines, allows the grapes to come into being and ripen of their own accord. His cultivation, then, does not reduce the grapes to the purpose for which he grows them. They are never simply "grapes for wine." He may use the ripe grapes to make wine, but the cultivation of the plants looks to their own "preservation and nurture" and not to their use. Construction does the same thing for what it builds. If I construct a jug, I may use it to pour wine, but the construction itself must look to the jug's own being if it is to be construction as a form of dwelling. The potter must think not only of the jug's use for him, but of its total context: the earth from which its clay has come, the sky whose rains provide the liquid that fills it, and the human community that will come together in drinking from it. Heidegger gets at the root activity of both cultivation and construction with the terms "freeing" and "sparing." We may be tempted to think of "sparing" as a fairy tale term, used of kings who spare prisoners brought to them. Or we may think of the phrase: "spare me the details," a request to cease from speaking. These uses describe a negative sparing, a refraining from action or speech. Heideggerian sparing, on the other hand, "is something *positive* and takes place when we leave something beforehand in its own essence, when we return it specifically to its essential being, when we 'free' it in the proper sense of the word into a preserve of peace,. . .the free sphere that safeguards each thing in its essence."[18] This sparing, or freeing, cannot be separated from the thing that it frees, since it actively supports the existence of what it frees. Without the freeing, nothing has an essence; things cease to be.

How does the freeing of the 1951 essay differ from the freeing of *Being and Time*? In *Being and Time*, I must give space for a thing in order for it to become a tool. I cannot use a tool unless I can first bring it close to myself and orient it toward a use, and I cannot bring it close to myself unless I allow it to be spatial in the first place. This means freeing it for spatiality. While this freeing is prior to any use I may have for the thing, the freeing is always ultimately for use: the turning of the thing into a tool. In other words, I do not have the thing's own best interests at heart. The freeing of the 1951 essay leaves the language of "use" and "tools" behind. The freeing is not simply prior to the orientation of tools for their use, it is higher and self-directed. If "the manner in which we humans *are* on the earth, is. . .dwelling," then we are most distinctively ourselves when freeing things to be themselves. We may make use of the things that we free, but our use finds itself subsumed in a much larger context. Heidegger's middle and late works explore this larger context, which never strays far from the sacred.

The Heideggerian Sacred Place

Heidegger intended his course on "Parmenides and Heraclitus," taught during the winter term of 1942 and 1943, to explore the thought of these two early Greek thinkers. The course seems to have progressed more slowly and in a different direction than he had intended, for the surviving notes and transcripts indicate that Heidegger discussed only a single fragment from Parmenides over the entire course of the term. Heidegger chose to focus not on Parmenides or Heraclitus but on the nature of truth. He presented truth in the course as though it were the Greek view of truth, but, as several scholars have pointed out and as he himself later acknowledged, his presentation of truth in the course had less to do with the historical Greeks than with a transcendent ideal.[19] Well into the course of the term, Heidegger undertook a commentary on the final passage of Plato's *Republic*, in which Socrates tells what has come to be known as "The Myth of Er." The title character, Er, dies in a battle, but after ten days comes alive again to tell his fellow citizens of what lies beyond the grave. He says that as soon as his soul was parted from his body it came to a "certain daimonic place" (τόπος τις δαιμόνιος).[20] The meaning of this obscure phrase leads Heidegger through an extensive discourse on the nature of the divine, the nature of place, and the relation between the two in this "daimonic" or "sacred" place.

Vision and the Sacred Place

The term "daimonic" is, of course, the source of the English word "demonic," but the meaning of the English word is severely restricted in comparison with the Greek. If we call something "demonic," we do not intend to compliment it, or to characterize it as something we would choose to have in our lives. The word "demon," from which we derive the word "demonic," fares even worse. If we do not think of the horned and tailed being of Michelangelo, the term suggests an even more frightening, inscrutable, but above all malevolent being. The Greek term lacks precisely this malevolent character, as well as the tails and horns, which characterize Greek satyrs rather than daimons. In what follows, I will use the admittedly unfamiliar spellings "daimon" and "daimonic" to describe the Greek being and its characteristics. Heidegger attempts to free us from the banality of the modern demon by referring to a passage in which Aristotle uses the term in a sense which is utterly incompatible with the horned satyr. Aristotle says of those who seek wisdom that they "know things that are excessive, and thus astounding, and thereby difficult, and hence in general daimonic—but also useless, for they are not seeking what is, according to straightforward popular opinion, good for man."[21] As Aristotle describes it, the term "daimonic" refers to what is in general excessive, astounding, and difficult. He opposes this object of human inquiry to the ordinary objects of our existence known to everyone, and illustrates the two with a reference to the story of Thales and the milkmaid. Thales, one of the legendary Seven Sages of Greece, goes for a walk one night to study the stars and, because he fails to pay attention to the road beneath his feet, steps off the road and falls into a well. A milkmaid who witnesses the incident finds it amusing, since it seems to demonstrate that the knowledge Thales possesses, which she does not, is far less valuable than the ordinary knowledge of how to get from one place to another, which she possesses and he does not.

Popular opinion does not recommend what is difficult because it is useless for achieving human goods. What goods are we talking about here? Aristotle explains elsewhere what most people think the good is: either pleasure, honor, wealth, or some such thing.[22] They seize on these as the good because such things are "obvious and manifest." Aristotle suggests that none of these three suggestions by itself measures up to the true good for a human being: pleasure, because it is common to other animals; honor, because we want it only after we have done something good; wealth, because no one chooses it for its own sake. Heidegger lingers instead on the fact that most people seek what is "obvious and manifest." He suggests that what can become present to a human being holds no difficulty "because he can always find, going from one being to the next, a way of escape from difficulty and an explanation."[23] Sensual pleasure holds no difficulty as a goal because it shifts from object to object. When I have exhausted the pleasure of a cup of coffee, I may move on to the pleasure of a walk in the woods. Nothing holds me down and demands an explanation of my experience from me. This is not to say that I, or any other pleasure seeker, do not deal with reality. Without reality, there could be no cup of coffee or woods to walk in. Pleasure seekers simply do not focus on that reality, and so they make their home only in beings, and not being. As Heidegger puts it: "because they always have being in sight (*Auge*), although not in focus (*Blick*), and only deal with, and calculate, and organize, beings, they ever find their way within beings and are there 'at home' and in their element."[24] Heidegger's more or less arbitrary distinction between "sight" and "focus" should not trouble us too much. What is really at stake here is the calculation and organization of beings. In *Being and Time* this is simply our natural activity as human beings, but now it constrains us "within the limits of beings, of the real, of the 'facts,' so highly acclaimed," where "everything is normal and ordinary."

The ordinary transforms itself when we get being in focus, and do not simply have it in sight. When we focus on being, "there what is not ordinary announces itself, the excessive that strays 'beyond' the ordinary, that which is not to be explained by explanations on the basis of beings. This is the extraordinary."[25] The very structure of the word "extraordinary" (*Ungeheure*) indicates that it is not the ordinary (*Geheure*). On this basis, we may be tempted to think of the ordinary and the extraordinary as two different classes of beings. One kind we experience every day—the coffee cup, the woods—while the other kind we experience rarely, if at all. To this class of being would belong the miraculous and the romantic: a budding leaf on a long dead tree or the sudden flash of a mountainous chasm through a train window. Aristotle's philosopher would in this case seek out such extraordinary experiences as his good. This treatment of the extraordinary has no place among Heidegger's Greeks. Their extraordinary is "the simple, the insignificant, ungraspable by the fangs of the will, withdrawing itself from all the artifices of calculation, because it surpasses all planning." Such an extraordinary cannot be sought out, since it stands outside the compass of human manipulation. Its independence of us does not mean that we do not encounter it, since it never ceases to present itself in the ordinary. Our inclination to differentiate the extraordinary from the ordinary is part and parcel of our parallel differentiation of secular from sacred, a differentiation which has become so ingrained in us that "for us it is difficult to attain the fundamental Greek experience, whereby the ordinary itself, and only insofar as it is the ordinary, is the extraordinary."[26] The Greek takes a jug of wine, for example, and pours out a libation to Bacchus, the god whose gift it is. With this libation in mind, we may lend some concreteness to Heidegger's somewhat oblique statement: "the extraordinary is that out of which all that is ordinary emerges, that in which all that is ordinary is suspended. . .and that into which everything ordinary falls back."[27] The wine poured out from the jug emerges from sources unseen by the Greek, but present in the context of myth, where they are identified with Bacchus. Bacchus lives on in the drink, suspending it in its existence, and its pouring out in libation returns it ultimately to him. The pouring out of a libation makes explicit the presence of the extraordinary even in the ordinary pouring out of wine for human consumption. In this way, no radical distinction of sacred and secular emerges. Only if we succeed in surmounting the distinction between secular and sacred may we identify the extraordinary with the daimonic. The daimonic is not the ordinary, but it "is what presents itself in the ordinary and takes up its abode therein."[28] The ordinary is what allows the daimonic to become present; it is the *place* of the daimonic. Heidegger derives the Greek term "daimonic" (δαιμόνιος) from δαίω, which means "to present oneself in the sense of pointing and showing."[29] The daimons are thus the "self-showing ones." They show themselves in the jug of wine because its ordinary use of pouring out drinks for human beings never wanders far from the extraordinary presence in it of the gods who are its source, ground, and end.

Since Heidegger gives "daimonic" a signifying function by relating it etymologically to "pointing," we are not surprised to find him discovering a similar relation in "divinity" (θεῖον). Heidegger finds in the Greek term "divinity" an etymological relation to the Greek term "to look" (θεάομαι). Heidegger does not deny that the usual sense of "looking" describes a relation between someone who looks and a thing looked at, but he notes that in Greek "to look" has only a medial form, which for Heidegger intimates an identity between the subject and the object of the verb. The grammatical structure of the verb indicates that our ordinary sense of "looking"—a looking *at* something—grounds itself in a deeper meaning: "looking, even human looking, is, originally experienced, not the grasping of something but the self-showing in view of which there first becomes possible a looking that grasps something."[30] The kind of looking that grasps something depends on a kind of looking that does not grasp anything, but reveals the one who looks. Before I can look at anything in the world, I have to be in the world myself. I have to "show myself," or "become present" in the world.

The very idea of showing myself already implies someone *to whom* I show myself. Otherwise nothing is shown. Someone who looks in this fundamental way does not acquire and understand an object, but "experiences the look, in unreflected letting-be-encountered, as the looking at him of the person who is encountering him." I must offer myself to be looked at, and so establish myself in a relation to something outside me. If this happens, "then the look of the encountering person shows itself as that in which someone awaits the other as counter, i.e., appears to the other and is." In other words, to look is to wait for someone else, and in waiting to appear to that other person. The one who looks is the one who appears, and this appearance does not reveal the mere surface of the one who looks. It cuts straight to the essence. As Heidegger puts it: "the looking that awaits the other and the human look thus experienced disclose the encountering person himself in the ground of his essence." I offer up my very essence in my look, and I can hardly do otherwise, since the look is nothing but a self-offering.

For Heidegger, the discussion of human looking here is but prologue to its extension to divine looking. The divine, too, reveals itself in its look: "that which looks into all that is ordinary, the extraordinary as showing itself in advance, is the originally looking one in the eminent sense: *τὸ θεᾶον*, i.e., *τὸ θεῖον*."[31] The divine defines itself—indeed, it gets its very name—by its possession of the most fundamental form of looking. This divine looking is no more a literal use of the eyes than is the most fundamental form of human looking. But the gods do have eyes: in the human forms taken by Greek statuary. The Greeks represent the gods in human form not because they think the gods are just powerful human beings, or because they "anthropomorphize" natural forces, but "because the Greeks experience man as the being whose being is determined through a relation of self-disclosing being itself to what, on the basis of this very relation, we call man."[32] Our being is determined through a relation between divine being and ourselves. In other words, we are not separate from the gods in the way that the statues of the gods are. The statues as beings do not call me with their look. The appearance of the statues, however, can through their relation to divine being—they are the gift of the gods—lead me to the apprehension of a more fundamental look. In this, "the looking performed by man in relation to the appearing look is already a response to the original look, which first elevates human looking into its essence."[33] The fundamental looking by which I show myself before the statue would never come about unless I were responding to the even more fundamental look of the gods—their showing of themselves through the statue. The look of the gods is nevertheless not separable from the human look in which it appears for the first time. This is why the gods appear in the statue in human form.

The Parmenides course makes clear to us two ways of looking, one which sees the secular and one which sees the sacred. Secular looking has being in sight as it grasps objects. Sacred looking has being in focus as it reveals the one who looks. We seem to make a distinction between secular and sacred only on the basis of the look we adopt toward things and not on the basis of any inherent sacred character present in the things we look at. And yet, Plato's myth of Er speaks specifically of a "sacred place." Heidegger reminds us that place is not "mere position in a manifold of points." It is the "originally gathering holding of what belongs together." He concludes: "a δαιμόνιος τόπος is an 'extraordinary locality.' That now means: a 'where' in whose squares and alleys the extraordinary shines explicitly and the essence of Being comes to presence in an eminent sense."[34] The gods do not appear to us directly. If they did, they would be beings and not being. For the gods to show themselves, they require a place, one where they can "shine explicitly." This place is the sacred precinct: the temple, the things that are consecrated there, and especially the statue of the god enclosed by the temple. It looks at us, but its looking does not grasp us, since it is made of stone. Its eyes receive no colors or shapes, and are not connected to a mind that could compose them into representations of objects. But the statue is more than mere stone. It is the place where the god "comes to presence in an eminent sense." To see why the statue, though made of stone, may still be the self-showing look of a god, we must investigate a second side of the sacred place, one that is not a work of sight but of speech.

Language and the Sacred Place

We do not simply await the presence of the gods in the sacred place. We also name them. As Heidegger describes it in the Parmenides course, the naming of the gods is another way of expressing the action of the place. The English word "God," as well as its German cognate, "Gott," derive from the same Indo-European root, a root Heidegger understands to mean "the invoked one." As we have seen, he takes the Greek names for "daimon" and "divine" to contain in their etymology quite different clues as to the nature of divinity. These names describe the actions of the gods, who "point" to themselves and who "look" in a way that reveals them. Where the name "God" refers only to what human beings do—they pray or invoke the divine—the Greek names refer to what the gods themselves do. They have then a more explicit tie to the extraordinary than the German and English names. The names of "pointing" and "looking" are not artifacts of language, which signify a content different from themselves. Naming itself is a form of pointing and looking, and so when we name the gods in Greek, we perform the action that the word describes. As Heidegger puts it: "the name and designation of the divinity (θεῖον) as the looking one and the one who shines into (θεᾶον) is not a mere vocal expression. The name as the first word lets what is designated appear in its primordial presence."[35] The name does not control the god, since the name cannot comprehend the god, but it does allow the appearance of the god, which puts a great deal of responsibility into human hands. The gods themselves do not make use of words. As Aristotle tells us, we human beings are the "animal that possesses the word" (ζῷον λόγονἔχον), and the foremost use of the word is "the letting appear of being by naming." Language is not primarily a means of communicating ideas from one mind to another. In its essence, it lets divine being become present. Language here takes up the role that Heidegger gave to place, in the *Sophist* course, seventeen years earlier. In the *Sophist* course, the place of a thing allows it to become properly present for the first time. A fish is properly present not when on a dinner plate, but immersed in water. In the Parmenides course, language allows the divinities to become present. This is not to say that language exhausts itself in the naming that allows divine being to become present. Much of our use of language directs itself at the manipulation of beings. Heidegger reserves a certain Greek term for the more essential use of language: *mythos* or "myth," which he renders in German as "saying" (*Sage*).

We have so far been speaking of the word as the spoken word, what a poet might speak in singing an epic poem. Heidegger warns his audience that this spoken word does not get at the word's essence. "The essence of saying does not consist in sound but in a tune in the sense of soundless attuning." The language that we use, and the stories that we tell, are the playing out of an attunement, which does not exhaust itself in an audible speaking. Language does more than spew sounds and mix concepts. As attunement, it brings "the essence of man to itself, bringing it, namely, into its historical determination." That is to say, language itself gives a people its identity as a historical people. The Parmenides course, here takes up a theme introduced by Heidegger in his 1935 essay "The Origin of the Work of Art," where he attributes the identity-giving power to the temple. In the Parmenides course, Heidegger has suggested that the statue of the god does not look at us in the sense of grasping because it is made of stone. Yet, it can reveal the god because it is the visible ground of the being properly housed in language. As Heidegger will say explicitly in his 1946 essay "Why Poets?", "language is the precinct (*templum*), that is, the house of being."[36] Heidegger's quip, "language is the house of being," has become famous, but it is rarely remembered what kind of house it is. Language is a temple that constitutes a world; the "Origin of the Work of Art" simply specifies that the intelligible temple of language is itself grounded in a visible temple.

In "The Origin of the Work of Art," Heidegger takes up the discussion of the temple as a seemingly innocent example intended to illustrate an entirely different point: the relation of art and truth. He tells us that he has chosen the temple simply because it "cannot be ranked as representational art."[37] Because representational art tries to depict something, we are tempted to judge its truth based simply on how accurately it has depicted its object. Such truth does not touch on the truth appropriate to art. The temple, however, is more than an innocent example of nonrepresentational art. The temple plays, literally, a central role in the context of the historical people that builds it. The people understands itself and its world only in relation to the temple. All other things appear in the light of the temple, just as the points on the circumference of a circle have meaning only in relation to its center. The role of the people's temple thus overlaps with the role of its language. Both condition the things that the people encounters as constituents of its world.

In the Parmenides course, Heidegger describes the daimonic place as one "where the essence of being comes to presence in an eminent sense." The temple, too, makes divinity present. At its most obvious and least significant, the temple simply encloses a statue of the divinity, which in turn looks out into the sacred precinct through the open portico. The sacred precinct would not be sacred unless the divinity consented to be present, and so its presence "is in itself the extension and delimitation of the precinct as a holy precinct." This does not make the temple the passive receptacle of divinity. The temple acts. It "fits together and at the same time gathers around itself the unity of those paths and relations in which birth and death, disaster and blessing, victory and disgrace, endurance and decline acquire the shape of destiny for human being." Here we see the temple taking up the role of place that Heidegger will give to it in his later essays: it gathers. At this point in his life, he sees that what must be gathered is not the less politically relevant fourfold of earth and sky, mortals and immortals, but the world of a historical people. The temple can open up a world by allowing all things to appear in relation to it. The easiest examples of this are those that reveal motion: a storm, for instance. The storm is in motion, but we do not see the violence of that motion unless we can set it in relation to something stationary. The temple serves as such a stationary thing: "standing there, the building holds its ground against the storm raging above it and so first makes the storm itself manifest in its violence." The temple does more than reveal motion; it reveals everything. Through the temple, "tree and grass, eagle and bull, snake and cricket first enter into their distinctive shapes and thus come to appear as what they are." The people only sees the eagle and oak tree in the context opened up by the temple. The eagle is the herald of Zeus; the oak tree is sacred to Zeus, and lets him appear in oracular form at Dodona. The temple that allows the god to appear also allows the things within the world to appear in all their proper relations to that god and to each other.

Heidegger's later Parmenides course does not flatly contradict "The Origin of the Work of Art." It goes on to articulate a relation between the temple that is language and the temple that is visible and architectural. The reconciliation begins with an objection posed by Heidegger to his own theory. He wonders why, if architecture is the place where the gods become present (as in "The Origin of the Work of Art"), the word can be the primary place of the gods' appearance: "if the divine essence appears precisely for the Greeks in the architecture of their temples and in the sculpture of their statues, what happens then to the asserted priority of the word?"[38] The introduction of sculpture and architecture brings the discussion into the realm of art, where it risks foundering on the modern development of aesthetics as a distinct discipline within the study of philosophy. Heidegger cautions us that we must not think "aesthetically" about the sculpture and temples of the Greeks. By "aesthetically," he explains that he means the consideration of the work "with regard to its effect on man and on his lived experience." Such consideration treats the work not as opening up a world, but as a mere object in space, which produces a feeling of pleasure in us when we contemplate it. This approach to the sacred place overlooks the world in which both the work and the human become properly present, and so it cannot see the necessary connection between the place and the word. Architecture, like sculpture, does not employ words as its material in the way poetry does. It also does not need the word in its capacity either to make sounds or to communicate information. It needs the "silent word" that constitutes attunement. This, as we have seen, is language as context for human existence. Without this attunement, "the looking god as sight of the statue and the features of its figure could never appear." The silent word, which provides a context for my existence, allows me to see the statue as more than stone. Without the word, I can take the statue of the god and use it as a doorstop, or a department store mannequin. Only the silent word allows me to see the look of the god in the eyes of the statue. The temple too needs the silent word, though it has no eyes in which the look could appear. The temple "could never, without standing in the disclosive domain of the word, present itself as the house of a god." "The Origin of the Work of Art," together with the later Parmenides course, present a coherent picture of the sacred place as the visible center of a world not directly visible. Only because there is a world of myth can the place be sacred, since only the myth allows a person to see the building as the house of a god.

The 1950s Lectures

In Heidegger's lectures of the early 1950s—"The Thing," "Building Dwelling Thinking," and "The Question Concerning Technology"—the gods have departed. The world-grounding event of National Socialism has not panned out, and the dream of a people united around a common heritage of myth and grounded in a distinctive place has disappeared from Heidegger's writing. Yet he has not abandoned the study of the sacred. He seems only to have adjusted slightly the degree to which we may expect the gods to be present. During the National Socialist period the gods had not yet arrived, but their arrival seemed imminent, in that a new world would be opened up for the German people.[39] Waiting for the gods could be expected to come to an end. Now waiting is an enduring and essential component of our relation to the gods, and of our own lives. Human beings "dwell in that they await the divinities as divinities," and they "do not mistake the signs of their absence." One thing human beings must not do is to "make their gods for themselves" and so "worship idols."[40] We cannot intentionally construct a world for ourselves. We can only wait for it to be given, an act that in these essays is "unhoped for." Waiting for absent gods does not deny the possibility of interaction with them. Places may still gather the gods into them, and so become sacred, but they gather the gods as absent. Heidegger thinks of gathering as the work of a place as early as the 1930s, but he fully develops his conception of it only in these lectures. He does not borrow this definition from Aristotle, who never describes place as a gathering. The Neoplatonists, both Hellenic and Christian, do make gathering essential to their conception of place, but Heidegger is unlikely to have relied on them here. He provides only an etymological reason for his use of gathering, explaining that the English word "thing" and the German word *Ding* both were used at one time to mean a gathering, so that a thing in essence is nothing other than a gathering. Two kinds of things can gather: things like the jug of the "Thing" lecture, and things which are also locales, like the bridge of the "Building Dwelling Thinking" lecture.

The jug, Heidegger explains in "The Thing," gathers in both its secular and its sacred uses. It gives us a gift. At times, this gift is for human beings, or mortals, as Heidegger calls us in this essay: "the gift of the pouring out is drink for mortals. It quenches their thirst. It refreshes their leisure. It enlivens their conviviality."[41] The pouring out performed by the jug need not be reduced to a single purpose. It does not simply fill glasses in order to quench thirst. The pouring forms part of a larger context of conviviality in which we find ourselves as human beings. If we reduce the context of conviviality to the mere pouring of drink to satisfy thirst, the jug ceases to be part of our identity as human beings, and we lose ourselves. In all the ways it participates in the life of human beings, the jug's pouring is secular, by which I mean only that it does not directly involve the divine. In the world of a historical people, the jug may also be used to pour out libations. Its gift may be "given for consecration. If the pouring is for consecration, then it does not still a thirst. It stills and elevates the celebration of the feast."[42]

The divinities of the consecrated libation dwell in the context of the feast. If we simply consecrate the libation mechanically, without participating in the context opened up by the feast, we lose sight of the gods.

Though consecration is no longer possible for us, who are not a historical people, the jug may still retain a sacred character. The jug can never be entirely responsible for its gift, since it does not produce the drink that it pours. It receives that drink as the prior gifts of rain from the sky and wellsprings from the earth. These prior gifts reunite the mortals and gods that are separated in the different purposes of the outpouring. When we consider only these prior gifts, we see "in the gift of the outpouring earth and sky, divinities and mortals dwell *together all at once*."[43] My pouring out the drink from the jug continues the gift that begins with the giving of the rain to the earth, and continues in the giving of the spring from the earth. These gifts are not mine. They bring my life together with something other than and beyond it. The spring is not simply the past gift to the jug. Instead, "the spring stays on in the water of the gift. In the spring the rock dwells, and in the rock dwells the dark slumber of the earth, which receives the rain and dew of the sky."[44] Though the spring and the other givers—the rock, earth, rain, and sky—do not manifest themselves to the eyes as the jug does, they are present in the gift through the drink that has been given through all of them. This presence is not the passive presence of something manipulable. The earth is not in the drink like a sugar cube, which can then be manipulated by the drinker—poured out at will or left to dissolve. The drink actively brings together earth and sky, gods and mortals. It "stays" them, to use Heidegger's term. He says: "staying is now no longer the mere persisting of something that is here. Staying appropriates. It brings the four into the light of their mutual belonging."[45] The jug gathers the four together. It is a human artifact, but without it, the water remains in the earth, and cannot be poured out in the context of conviviality. We may still just cup our hands together and drink straight from the spring, and forgo the jug altogether, but the mere sucking up of water from the spring inhibits, rather than enlivens, the conviviality of human beings. We bend over to face the water, and must drink quickly, so that it does not slip through our fingers. The jug, then, does not just allow the storage of water or wine. It gathers us together in a community. The gods are present in this community as the givers of the gift, but they remain absent because the communal context has no names for them. We treat the jug as sacred in this context not by consecrating it, but by freeing it so that it may gather the earth, sky, immortals, and mortals. We must let the wine be a gift, without laying claim to its giver.

Certain things differ from the jug in structure. They do not simply gather the fourfold of earth and sky, divinities and mortals; they gather it in a way that allows an "area" (*Stätte*) for it. Such a thing can accomplish this "making space for an area" because it is itself a "locale" (*Ort*). A locale, first and foremost, means a thing that we are able to move through, like a room or a hallway. But there is another difference between the locale and other things. The locale does not contain the gift of the fourfold within itself, as the jug does. Instead, it serves as a kind of nexus around which the four can appear. Heidegger chooses a bridge crossing a stream as an example of such a nexus. The bridge, first of all, is not added to existing banks that stretch indefinitely in either direction along the water. We do not consider the banks as banks until the bridge brings them together by crossing the stream. As Heidegger puts it, the bridge "brings stream and bank and land into each other's neighborhood."[46] It "*gathers* the earth as landscape around the stream," and so allows the earth to appear for the first time. The bridge gathers the sky as well as the earth both by being constructed to handle the floods that result from a downpour, and by covering and uncovering the waters that pass beneath it so as to construct an explicit relationship with the sky. Perhaps easiest to see is that the bridge gathers mortals. It may connect castle and cathedral square, town and village, village and field, initiating the motion of mortals in relation to each other. The crossing over of the bridge is also a symbol of our crossing over as mortals from life to death, and so the bridge gathers the divinities simply by being "a passage that crosses." Not everyone recognizes the presence of the divinities in the bridge, and when they do, their recognition does not always take the same form. Heidegger gives only one example: the saint of the bridge, whose statue or at least implicit presence places the act of travel under the auspices of the divinity. In the absence of a historical people, the bridge may still symbolize the crossing from life to death, but this crossing can no longer be enfolded in the mythic context of a crossing to Valhalla or the Elysian fields.

The bridge, like the jug, gathers the fourfold, but unlike the jug, it does not contain in its gift the fruition of a series of gifts: from the sky, then the rain, then the earth, rock, and spring. The bridge does not contain at all, and the language of gift has disappeared. In the jug, the giver remains in the gift. In the bridge, the giver has not *given* a gift. The bridge lets the earth appear as the landscape. It lets the immortals appear in the symbol of crossing. If the language of gift remains relevant at all, it is in the gift of this appearance. Since every place allows the immortals to appear, every place has a sacred character. These places are secular when compared with the consecrated places—the jug used for libations, or the temple precinct—but they are sacred in relation to the degraded secularity which takes no account of the gathering performed by things, and so reduces them to single purposes.[47] In the realm of this latter secularity, the pouring of drink from a jug devolves into the pouring of drink from a tap in a bar. The source and container of the drink recede in favor of a more efficient accomplishment of pouring, the single purpose of the tap.

Naturally Sacred Places?

The sacred place can never come about naturally—meaning independently of the life of a historical people—at least within the context of Heidegger's "Origin of the Work of Art" and Parmenides course. Heidegger's temple gathers together a context, a web of meanings, which is not the same for every human being, but is rather "the world of this historical people."[48] In the case of the temple at Paestum described in "The Origin of the Work of Art," the ancient Greeks are the historical people. The temple at Paestum gathers no context for a visiting Egyptian, whose context is provided by his own temples in his own land. In Heidegger's terms, the Egyptian sees the temple not as a work, but an object, which gathers no world for him. The Egyptian may acknowledge that there is a god in the temple, but it is the god of the Greeks, not of the Egyptians. Once the historical people disappears, the temple ceases to be a work, and, more important for our purposes, it ceases to be sacred. The god withdraws from the temple.[49] When we, who are not ancient Greeks, visit the temple at Paestum, we see only an object, a bygone work. It belongs now to the realm of "tradition and conservation," not the sacred. The sacred, then, is essentially public, because its character depends on its ability to open up a context within which the life of the historical people that built it becomes meaningful. If my people does not worship the Olympian gods, then the context of meaning in my life cannot derive from the temple of Hera at Paestum. I cannot hope privately to establish a belief in Hera as the mother of the Olympian gods, and so experience the temple at Paestum as sacred when I visit it. For this reason, too, the temple can never be a naturally sacred place. Such a place would have some effect on all human beings, not just members of a given historical people.

The historical record presents some troubling cases for a straightforward reading of the Heideggerian world-grounding sacred place. In the year 527, for example, the royal publicist Cassiodorus wrote on behalf of the Gothic king Athalaric to order the elimination of violent robberies that were taking place at the sacred place of Marcellianum during its great fair, held every year on the feast day of Saint Cyprian. The place was considered holy because of a spring whose waters rose miraculously during the baptismal rites. Cassiodorus gives a description of the place full of pastoral imagery, noting its "delightful meadows" and natural cave, from which "pours forth a fluid of such clarity that you would suppose to be empty a pool which you know is brimful." The pool contains a school of sacred fish, who "come boldly to the hands of those who feed them, as if they know they are not to be caught: for whoever dares to do such a deed is swiftly to feel the vengeance of the deity." The spring is named Leucothea, which in Greek means "white." Cassiodorus conjectures that the spring is so named because of "the clarity and great whiteness of the water in that place." S.J.B. Barnish suggests a better reason for calling the spring "Leucothea."[50] It may have once been a shrine devoted to the Hellenic goddess of the same name, whose rites had been widespread within the Mediterranean basin. Barnish notes that Leucothea is identified with the Roman goddess Mater Matuta and the Syrian goddess Atargatis, who was saved from drowning by fish, and so kept sacred fish in pools. There is no evidence that Marcellianum ever had a Hellenic shrine to Leucothea here, but if there were one, the crossover from Hellenic to Christian rite would have required little effort. Barnish notes the continuity of symbolism between the two rites: "like the goddess, catechumens were submerged in water to attain immortality; as she was saved by a fish, so they were by Christ, of whom the fish is a familiar type; like the children in the cult of Matuta, one of whose functions was as a goddess of childbirth, they were drawn into a wider family."[51] It may be that some powerful cleric or lay Christian consciously replaced the Hellenic rite with a Christian rite at Leucothea, and so allowed the symbols of the Hellenic rite to persist, albeit in a changed world.

Leucothea is not the only sacred place to provide evidence of a conscious attempt to maintain the continuity of holy landscapes in Christian Europe.[52] In these places, the symbolic content of the natural objects, and so also the rites which visibly ground that content, remain after the advent of Christianity, but transformed in relation to the divinity whose power invests them. Because the symbols remain in the Christian rite at Leucothea, Barnish may conclude that the founder of the Christian rite was not eliminating the naturally sacred character of the place, but "maintaining its sanctity by associating it intimately with the one God."[53] What could provoke a landowner to maintain the sanctity of the spring at Leucothea? A desire to preserve one of the centers of a people, which remained largely unchanged by its conversion to Christianity? Or did the landowner recognize that the place itself was sacred to the gods—perhaps to any gods—and so needed to be protected and revered? Does the sacred character of the place depend on the people, or does the people depend on the sacred character of the place?

Eliade: " 'Our World' is Always Situated at the Center"

Mircea Eliade, who did as much as anyone in the twentieth century to establish the study of sacred place as a serious field of anthropological research, has become a controversial figure for claiming that the sacred place constitutes the "center of the world."[54] Eliade's sacred place originates when something divine or wholly other manifests itself to a particular person or people. As reported by his critics, Eliade locates the power of the divine in the place itself, and not in any relation between the place and the people who hold it sacred. John Eade puts it this way: "the power of a miraculous shrine is seen to derive solely from its inherent capacity to exert a devotional magnetism over pilgrims from far and wide, and to exude of itself potent meanings and significances for its worshippers."[55] The shrine may give meaning to a particular people, but it seems incapable of dying out when that people disappears. Possessed of "inherent capacity" to draw pilgrims, it may remain dormant for some time, but is always capable of attracting new individuals or new peoples to whom it can give meaning.

Eliade's treatment of the sacred place in his 1957 study, *The Sacred and the Profane*, gives a slightly more nuanced reading than that of his critics. The sacredness of a place, as Eliade describes it, does not arise from qualities intrinsic to the place itself. It requires some special interaction between a human being and the gods. This may be a theophany or hierophany, an actual appearance of a god, but it need not be so dramatic. A sign of the god may appear—something that "does not belong to this world." Eliade tells of the founder of El-Hamel, who settled in the place because his staff budded there when he thrust it into the ground for the night. In the absence of a theophany or an unlooked-for sign, a people that needs a sacred place may intentionally provoke a sign. A wild animal may be hunted and a sanctuary built on the spot where it is killed. In any case, "the *sign*, fraught with religious meaning, introduces an absolute element and puts an end to relativity and confusion."[56] We find here a pragmatic approach to religious practice to be echoed later by Walter Burkert, who will comment on the Greek art of divination: "the aid to decision-making, the gain in self-confidence, is more important than real foreknowledge."[57] The religious practice serves as a tool to produce a feeling of confidence in the one who uses it. We can imagine other examples of theophanies and signs that depend on some characteristic intrinsic to the place—a clearing in the woods, an unusual rock formation, a powerful waterfall, or a quiet hollow in the hills. It is nevertheless telling that Eliade's examples include nothing like this. He wishes to stress that the sacred place is determined by rites that have meaning only within a human community.

The characteristics of Eliade's sacred place once it has been established bring it slightly more in line with the representation of his thought provided by his critics. The founding of the sacred place reproduces the act of creation, and so the sacred place itself takes on the characteristics of the total cosmos. It is, in Eliade's words, an *imago mundi*, or "image of the world." We need not assume with Eliade's critics that the world here is the same for every people. The temple of Hera may be the center of the world for a Greek of the fifth century before Christ, but a contemporary Egyptian may find it in the temple of Neith in Saïs.[58] In certain cases these centers may overlap, as at the church of St. George near Beit Jala, which until recently was visited by Christians (venerating St. George), Muslims (venerating al-Khadr), and Jews (venerating Elias).[59] Eliade does not address this superimposition of centers and confrontation of worlds, but he does note that, even within the world of a single historical people, there will be many centers. They will have different sizes, so that some will contain others. A sacred city, for instance, may be itself the center of the world, while containing in it several shrines, each of which is also the center of the world. Residents of the city will themselves desire to make their homes into centers of the world as well, a task they accomplish by building each home around a symbolic center—a central supporting pole or, later, a hearth—or by ritually sacrificing an animal symbolic of the primordial animal from which the world was made. The multiplicity of centers poses no contradiction, since, as Eliade has already made clear, we are not speaking of a center in a homogeneous geometrical space. He says: "the multiplicity . . . of centers of the world raises no difficulty for religious thought. For it is not a matter of geometrical space, but of an existential and sacred space that has an entirely different structure."[60] If Eliade means that the religious person thinks of the multiplicity of centers as extending beyond his world, then we find ourselves with the problem of naturally sacred places. If, on the other hand, the religious person sees many centers only within his own world, outside of which there are no centers, then we must say that Eliade is subtler than the representation of his thought posed by his critics, though we now find ourselves with the problem of explaining the interaction between different peoples.

Eliade's important discussion of symbolism only complicates the matter. He implicitly raises the question of interaction between the worlds of different historical peoples when he suggests that some symbols are universal, a claim that seems to leave him open to his critics. A symbol that is universal must be rooted in nature and not in a particular historical period or people. Yet, Eliade does not suggest that we encounter universal symbols *as* universal; they are handed down from one religion to another.[61] His primary example here is the handing down of symbols from the pre-Christian rites to their Christian successors. Eliade does not mention any specific pre-Christian rite, presumably because he wishes to treat pre-Christian religion as primordial and ahistorical, but we may look at the specific case of the Hellenic rites. These often have, as in the case of Leucothea, a purification rite involving water. The purifying symbol of water remains in the Christian rite of baptism, a fact which leads Eliade to conclude: "the revelation brought by faith did not destroy the pre-Christian meanings of symbols; it simply added a new value to them."[62] This adding of value, admittedly a modern expression that likely does not characterize the approach of the early Christians to their rites, nevertheless reveals that Eliade does not think that the symbol plays the same role throughout history. He says: "history constantly adds new meanings, but they do not destroy the structure of the symbol."[63] It seems, then, that the universal symbol finds expression only in the particular context of a historical people.

Victor Turner: "The Communitas Spirit Presses Always to Universality"

Eliade's work on sacred place had a great deal of influence on the anthropologist Victor Turner, whose theory of the sacred place may be found in *Dramas, Fields, and Metaphors* (1974) and *Image and Pilgrimage in Christian Culture* (1978), the latter written with his wife, Edith Turner. Victor Turner does not parrot Eliade. He does adopt Eliade's account to explain the believer's view of how his particular pilgrimage center came about. A pilgrimage center appears where "according to believers, some manifestation of divine or supernatural power had occurred, what Mircea Eliade would call a 'theophany.'"[64] Turner also notes that, once the center has been established, it generates a "field," by which Turner means that the center becomes the stimulus for the building of "cities, markets, and roads."[65] Pilgrims need access to the center so roads are built. They need lodging and food, so inns and markets appear. A class of permanent residents lives in the newly founded town, which develops its own civic structure. This "field" is not the "world" of Eliade, since it does not have the pilgrimage center as its own center. The residents of the town may think of the pilgrimage center simply as a revenue source, while they belong to a different cultural context with its own center. Should the pilgrimage center ever become the center of a world, it would cease to be "that form of institutionalized or symbolic anti-structure (or perhaps meta-structure) which succeeds the major initiation rites of puberty in tribal societies as the dominant historical form."[66] Pilgrimage center as anti-structure: this already suggests Turner's departure from Eliade and the Heidegger of "The Origin of the Work of Art." Structure, as Turner describes it, is the relation between different classes and roles within a society that we see manifested in "legal and political norms."[67] Eliade and Heidegger both suggest that the sacred place establishes these norms; it does not violate them. Turner's pilgrimage center decays when it becomes the center of a world, because it then becomes the ground of structure, rather than anti-structure. Its only hope must be that a new world will replace the world it has established, so that it may cease to be the center and rediscover its proper place on the periphery.[68] Turner does not deny that some sacred places ground legal and political norms, but he opposes these sacred places to pilgrimage centers, for example, in his discussion of fertility shrines and earth shrines in Africa. The fertility shrines are located at the physical center of the political community, and undergird the legitimacy of that community. The earth shrines are often established far from the political centers, and evince a partial dismantling of the dominant political structures. The Eliade-style fertility shrine and the Turner-style earth shrine establish, literally, different places.

The peripheral, or the liminal, as Turner prefers to call it: this is the place of the pilgrimage center. It could not stand farther away from Eliade's sacred place as "center of the world." The pilgrimage center has the liminal as its proper place because its essential character is what Turner calls "communitas," borrowing a term from Paul Goodman.[69] The real origin of the sacred place, which may or may not constitute a theophany, is an experience of unity with all other people, what Turner calls "spontaneous communitas." The attempt to capture and preserve this experience requires that it be rooted in a place. Regular visits to the place build up habits, and eventually social structures that may replicate those of the historical people from whom the sacred place initially appeared as an escape. But the sacred place, if it remains sacred, never becomes the mere ape of the historical people. It possesses what Turner calls "normative communitas," a structure which aims always at universality, the transcending of the historical people. The pilgrimage center may not be a naturally sacred place, but it strives to become natural by becoming universal.

Anthropologists have mounted a formidable critique of Turner's theory over the years since he first formulated it. John Eade and Michael J. Sallnow have gathered together a number of these critical readings in *Contesting the Sacred: the Anthropology of Christian Pilgrimage* (1991). After citing a number of studies which attempt to test Turner's theory, they note: "in none of these cases did the investigator find support for the theory; to the contrary, a recurrent theme throughout the literature is the maintenance and, in many instances, the reinforcement of social boundaries and distinctions in the pilgrimage context, rather than their attenuation or dissolution."[70] The resulting view of pilgrimage that plays out in *Contesting the Sacred* is that of a place where different classes come together not to experience communitas, but to embody a struggle for legitimacy and power. The meaning of the place results from this struggle, and changes with the make-up of the groups that visit it. This presentation of the sacred place matches up with neither Turner nor Eliade, and seems to eliminate the sacred character of the place altogether. The pilgrimage center becomes the kind of political place where different people interact in a way that recognizes their differences.[71] Not every anthropologist has adopted their critique. Simon Coleman and

John Elsner have suggested that it cannot explain why pilgrimage centers seem different from other social places: "while virtually all social practices are open to contestation, not all have the look of pilgrimage. In other words, the emphasis on the idea of pilgrimage sites being void of intrinsic meaning does tend to ignore the considerable *structural* similarities in pilgrimage practices within and between traditions."[72] Eade has acknowledged the force of this critique and, in the 2000 reissue of *Contesting the Sacred*, praises the "strengths of an eclectic approach."[73]

The Role of the Body in the Sacred Place

The contemporary emphasis on the sacred place as either establishing a world, or providing a place where a world can be contested, risks losing sight of the material character of places. This lack of attention to the body in its interaction with place is at least as old as Heidegger, who does not explicitly discuss the role our bodies play in the establishment and the experience of place. When he discusses the explicitly sacred place in the form of the Greek temple, Heidegger withdraws even further from a discussion of the body, as we have seen. He finds the sacred quality of the temple to be so far from its visible structure that he can identify it with the afterlife in Plato's myth of Er. Language is the true temple of being, though we may not separate that true temple from the temple we see, just as we may not draw a radical line between the sacred and the secular. We may well ask, however, if Heidegger has adequately accounted for the bodily nature of place, and specifically the sacred place, when he explains it in terms of language.

Heidegger and Edward Casey

The term "near" (*Nähe*) plays an important role in Heidegger's theory of place from its early exposition in *Being and Time* to its later development in the 1950s essays. We typically use the term "near" to describe what is within reach. An *object* within reach is one I can get at with a simple change of position or at least without much effort. A *place* within reach is one that requires little effort to arrive at. If someone asks me for directions on the street, I might respond: "it's nearby, just around the corner." In both cases, the term makes sense only with reference to the human body.

Heidegger's own use of "near" reveals just how little the human body matters to his theory. In *Being and Time* he says that, by bringing things near, we make them ready for use. In fact, he identifies the two acts: "the things at hand of everyday association have the character of *nearness*."[74] Some things are at hand only when I can reach them with my hands or feet. Others, like most things I hear about on the radio, come to be at hand even while remaining thousands of miles away. I turn on the radio and I hear, say, a story about an event in Afghanistan.[75] The people involved, the land described: these come to be near to me as I hear about them, while the radio in front of me becomes farther away. Nearness, then, is not keyed to anything bodily. All that matters is that the object be at hand in its own way. A wrench is at hand only when my hand can reach it, but Afghanistan may be at hand while remaining thousands of miles away. Heidegger's way of speaking about nearness here allows the possibility of a virtual world, in which we no longer use bodies at all, but see what electronic pulses present to our synapses, and manipulate electronic "objects" with complete power and freedom. Were we able to live entirely within such a system, we could then handle our "everyday association" without any body at all.

Heidegger's 1958 essay, "The Nature of Language," shows just how differently he later came to conceive "nearness." In *Being and Time*, "nearness" means ready for use. In "The Nature of Language," things that are ready for use may not be near at all. In this essay, Heidegger relies on a new term—"neighborhood"—which describes the result of nearness. Things which are physically present to one another may not be near each other, and so may not be in the same neighborhood. He says: "two isolated farmsteads—if any such are left—separated by an hour's walk across the fields can be the best of neighbors, while two townhouses, facing each other across the street or even sharing a common wall, know no neighborhood."[76] The contrast between the ancient and the modern here is meant to strike us. The modern townhouses may be stacked up, practically on top of one another, out of a concern to make everything ready for use, but they cannot produce the neighborhood of the scattered farmhouses. Certain forms of making ready for use explicitly destroy nearness. Heidegger suggests that the launch of Sputnik in 1957, the year before his publication of this essay, is the first salvo in a battle for control of the earth. The satellite, by taking up a position beyond the earth, is able to make virtually any position on earth ready for use. This transformation of the earth into readiness for use "is making a desert of the encounter of the world's fourfold—it is the refusal of nearness."[77] By making them ready for use, the satellite makes things on earth farther away.

We may well ask what brings things near, now that the act of making them ready for use does not. Heidegger has an answer, though he uses the slightly broader term neighborliness rather than nearness here. Neighborliness, he says, is "to be face-to-face with one another."[78] He does not mean relative proximity in position. The two townhouses in the above example have proximity in position, but they are not face-to-face with one another. Nor is the face-to-face encounter restricted to, or even primarily undertaken by, human beings. Heidegger notes: "we tend to think of face-to-face encounter exclusively as a relation between human beings," when in fact "it originates in that distance where earth and sky, the god and man reach one another." The mention of earth and sky, god and human, should immediately recall for us the fourfold described by Heidegger in earlier essays. The face-to-face encounter occurs between the four, and is enabled by the gathering of the fourfold when things and locales act as places. The jug allows me, as a mortal, to become near to the gods, the sky, and the earth, through my attention to its gathering.

When Heidegger says that neighbors face each other, he does not mean that they see each other's bodies. When the "face-to-face encounter" of the fourfold occurs, the four are present in a way that does not directly involve the body. I do not necessarily see the immortals—who remain absent—or the earth and sky, which in the case of the jug are present only in the wine they have given. Heidegger never seems to have found a form of nearness determined by bodily presence, neither in his early nor his late works, and so his treatment of place remains insecurely anchored in the body.

Edward Casey, who depends on Heidegger for much of his distinction between place and space, recognizes that Heidegger does not do enough to ground the existence of place in the activity and nature of the human body.[79] This is not just Heidegger's problem. Casey notes: "in ancient and modern Western philosophy there is rarely any serious discussion of the role of body in the determination of place."[80] By body here Casey does not mean an inert physical mass, but the living human body that perceives and moves around in the world. He finds in Aristotle alone a recognition of the mutual dependence of body and place, for Aristotle claims we only distinguish the dimensions of up, down, right and left by moving our bodies around.[81] For instance, we have to turn our bodies around before we can find out that "left" is not a quality in the things that happen to be on our left at a particular time. Aristotle does not privilege this relationship in his discussion of place, and it quickly disappears from the philosophical tradition. Casey does not wish to abandon both Aristotle and the tradition because of this deficiency. Like Heidegger, he wishes to develop Aristotle's insight that place has power, though Casey relocates the power of place to the human body: "a considerable portion of this power is taken *on loan*, as it were, from the body that lives and moves in it. For a lived body energizes a place by its own idiosyncratic dynamism."[82] Neither Aristotle nor Heidegger talk about the power of place this way, and so Casey chooses new muses when he wishes to discuss it: philosophers such as Immanuel Kant, Alfred North Whitehead, Maurice Merleau-Ponty, and Edmund Husserl, who discuss the importance of body in all human experience.[83]

None of these philosophers works through a distinction between place and space, but their work on the body is crucial, for only if the body and place are considered together "does the importance of place in distinction to space become fully evident."[84]

Places appear wherever a human being can gather a nexus of things. The office, the bedroom, the playground, the clearing: all these can be places because we can take them in within a single frame of reference for our bodily activity. We often speak of much smaller places, such as a cardboard box that a child has appropriated as a playhouse. This kind of a place comes quite close to the limits of the body itself, as I find out when I try to climb into the same box, and cannot fit. My body has prevented me from occupying the box as a place. The innermost limit of place seems to be the surface of the body itself or, as Aristotle puts it, the limit of the surrounding body, though we do not nowadays typically characterize such a surrounding body as a place. If I put on a shirt, for instance, I do not consider it a place. Place has an outermost limit as well, beyond which I cannot gather a group of things into a single nexus. When I say that I am in Dallas, for instance, I do not mean the same thing by "in" as I do when I say that I am in my office. To be in my office means to have within reach a number of things that I need: the telephone, the computer, the desk. The office gathers these things together in a way that puts them at my disposal. When I say that I am in Dallas, on the other hand, I refer to a region rather than a place, a region perhaps symbolized by the landscape visible as the city skyline. The region resembles the landscape by extending beyond what is within reach. The landscape, however, remains visible, while the region allows us to move from place to place without itself becoming visible. If it is to appear at all, it requires a symbol like a landmark or a landscape. Casey identifies a landscape as what I can see at once from a certain point of view, possibly containing many places within it, while itself still being visible. As he puts it, "a landscape seems to exceed the usual parameters of place by continuing without apparent end; nothing contains it, while it contains everything, including discrete places, in its environing embrace."[85] Our means of distinguishing a place from both a landscape and this kind of region is the same: the human body. To say that I am in Dallas is true, but does not say very much, because I experience only what is accessible to my body and its senses. A more meaningful answer indicates not just what I can see, but what I can experience through bodily interaction: not just the skyline from my window, but the things in my office.

The dependence on the body for the distinction between landscape and place already hints at the mutual relation between place and the human body. My office is a place and not a landscape because it gathers for my use all the things I need to write this book: the computer, a shelf of books, a chair, and a desk. All of these things are within arm's reach. Other places may be defined by my feet. What I can get to by a short walk also constitutes a place: the mall outside my office window, for instance, which extends from my office to the various classroom buildings of the university. My ability to reach with my arms or by a short walk with my legs, then, seems to determine what I mean by a place. Through my office window, I also have a glimpse of the cross timbers landscape: an expanse of various kinds of oaks, elms, mesquites, red cedars, and various other plants and animals that I cannot see but know are there. Were I a bodiless perceiver, occupying only a given point in space, I could not distinguish the character of my office from the character of the cross timbers. I would have no arms and legs to distinguish the scope I could give to a place. Casey does not conclude from this seeming dependence of place on the body that the body creates place. He appeals to "certain dimensional structures" that "inhere in things and places themselves and may reflect little if any influence from the incursion of human bodies into their midst."[86] When I put a book on the table, for instance, it will remain on the table even if I disappear. The book and the table have a dimensional structure, which allows relations of above and beneath, whether or not a human body is around to add its distinctive contribution. Though Casey is right to say that the book stays on my desk when I leave, he seems here momentarily to abandon the deeper structure that the human body gives to place, a structure identified and developed by Heidegger. Place is more than a dimensional structure. It requires a human being actively freeing the place to gather the earth into itself. We may still agree with Casey that body does not create place, but only because place emerges from an earth not produced by the human body. If the human being gathers a place from the earth, there must be an earth to be gathered. This is not to say that the earth is a previously existent "thing" or "place," which we simply set in order. We have difficulty expressing how the earth exists at all, since it lacks the definition that we give to things and places. Places, then, may depend for their definition on the embodied activities of human beings, but both human beings and their places are equally dependent on the earth, which is gathered into these places.

Lindsay Jones and Henri Lefebvre

We have seen that Heidegger treats the nature of place as somehow between body and language. His student and colleague, Hans-Georg Gadamer, takes greater account of the body in the experience of the particular kind of place that humans construct for themselves. He tells us in *The Relevance of the Beautiful* (1977) that, in order to distinguish a building from other art forms, "we. . .have to go beyond the purely artistic quality of the building considered as an image and actually approach it as architectural art in its own right. To do that, we have to go up to the building and wander round it, both inside and out."[87] We use our bodies to wander around buildings. We may imagine a disembodied perspective that is able to see the building from many different angles, but this disembodied perspective cannot get past the building "considered as an image." It can only stack up image upon image, each acquired from a different angle. The human body, on the other hand, has itself as a standard of measurement. It has its own standards of time and distance, which may or may not be respected by the place. Gadamer acknowledges architecture's character as establishing a place, with its bodily standard of measurement, but his work has paradoxically become implicated in the twentieth-century penchant for reducing all things, including places, to constructs of language. The reductive character of theories which treat all objects as texts, including places, has made hermeneutic approaches to architecture controversial in recent years, particularly among those who wish to establish the body as primary for being in place. Lindsay Jones' recent two-volume study *The Hermeneutics of Sacred Architecture* (2000) takes up the hermeneutic approach with an explicit awareness of its controversial nature. He builds up his work from a whole host of the brightest lights in twentieth-century continental thought, but he reserves the foremost place for Gadamer. This is not the Gadamer who distinguishes architecture as irreducible to an image, but the father of hermeneutics, who turns texts from objects with a stable meaning into interactions whose meaning depends as much on the reader as the text. Jones applies the hermeneutical method to buildings, which "mean nothing in and of themselves." They acquire meaning only in the "dynamic and fluctuating interactions between people and buildings."[88] Jones does not want to abandon the building as object altogether; he simply does not wish to focus on the building without reference to the human beings who perform their rituals within it. Rather than architecture, Jones wishes to focus our attention on the "ritual-architectural event." A troubling and perhaps unavoidable consequence of turning our attention to an event rather than a building is that we lose sight of the building's materiality, dwelling instead on the more "dynamic and fluctuating" realm of actions and interpretations. Jones reveals himself to be aware of the danger here, but he does not allow it to dissuade him from his project. He cites the work of Lawrence Sullivan, who, according to Jones, recommends that problems with treating objects as texts be overcome by "a widening rather than an abandonment of the hermeneutical project." Respectful treatment is now to be given to non-textual sources of meaning, such as "dreams, sounds and smells, light and shadow, table manners, bodies and bodily processes such as weeping and bathroom etiquette, along with pottery, basketry, canoe-making, and, of course, art and architecture."[89] Jones notes that Sullivan's recommendation "nearly provides a rationale" for his entire book. The trouble here is that Jones and Sullivan still treat these objects, activities, and sensations as bearers of meaning. The scholar still asks what sounds, smells, and lights mean within the church, for example, and not how they play a role in establishing the building as a place.

Jones fails to escape this temptation to treat the building only as a text in the course of summarizing the work of Paul Ricoeur, who seems to take the opposite approach to the building when he says: "the specific occurrence of a ritual might profitably be imagined as a text, but a specific building ought not be."[90] Jones' rationale for following Ricoeur here is that if the building were a text it would be a fixed text, since the building rarely changes. It would reveal a meaning or set of meanings that would themselves be fixed. We could say the same for any object, and this is where the problem emerges in Jones' analysis. He rejects an interpretation of the building in favor of an interpretation of the ritual event that takes place there, because the event, in Ricoeur's words, "appears and disappears."[91] It is thus better able to produce the "superabundance" of meanings that Jones desires. But a multiplication of the meanings of the building, to the point that it means different things even to different people participating in the same rite, does not eliminate the problems associated with treating a building as an object, a bearer of meaning. The multiplication of meanings simply prevents the building from having a universal or natural meaning, there to be revealed to anyone who properly interprets it.

Henri Lefebvre, in his 1974 work, *The Production of Space*, attacks the very notion that a work of architecture may be profitably treated as a text. First of all, the "reading" of a place does not adequately account for the role of the body in experiencing place. When someone encounters a new place for the first time, "he first experiences it through every part of his body—through his senses of smell and taste, as (provided he does not limit this by remaining in his car) through his legs and feet."[92] Our first experience of a place depends on what we sense—colors, sounds, smells, tastes, temperatures, and textures—and not what we interpret in the signs and symbols yielded by the place. Besides housing our sense organs, our bodies contribute to the experience of place through their shape and function. These provide us with their own distinctive mode of experience, quite apart from their ability to limit and condition the kind of meanings we get from the place. A plaza may trigger the same interpretation in me whether it is fifty feet or five miles wide, but my body experiences the two in quite different ways. Lefebvre concludes that the senses and the body have priority in our experience of places: "it is by means of the body that space is perceived, lived—and produced."[93]

Lefebvre also poses a more sinister problem with reading places as texts: this method is blind to the role of power in shaping the place.[94] Power stands outside of interpretation: it "can in no wise be decoded." It stands outside the realm of codes because it does not seek to communicate or to understand, but wants only to preserve itself, which it does through the reduction of all space to coded space. When the interpreter goes along with this, he lets himself in for disaster, because this puts all space in the service of the state, which "has control of all existing codes." Lefebvre's state is not the product of the conscious decision of a few people in power. It is the Heideggerian and Eliadean world, which "has control of all existing codes" because it determines what each thing will mean to the historical people whose world it is.[95] In one respect, Lefebvre's state differs from Heidegger's world: Heidegger's world is constituted by language. Lefebvre's state uses language as a tool to maintain power by establishing and controlling meaning. It grounds meaning initially, but also shifts meanings to maintain its power by taking advantage of a weakness inherent in language. There is "a mesmerizing difference, a deceptive gap," between the sign present in the language and the thing it hopes to signify.[96] Because sign differs from signified, the state can shift the relation of sign to signified so as to change the meaning of the signified. Lefebvre does not conclude with George Bataille that this weakness should lead us to subvert or abandon language, but he does stress the need to prevent our experience of place from reduction to reading or decoding.

Preeminent among the body's forms of resistance to coding is the sense of smell. We have already seen Jones cite smell as one of the non-textual bearers of meaning, but the fact that it bears meaning at all demonstrates how far Jones is from Lefebvre. If smells were primarily bearers of meaning, they could not function in the more basic role Lefebvre assigns to them. For Lefebvre, space itself, insofar as it appears to the senses, has "a basis or foundation, a ground or background, in the olfactory realm."[97] This is not to say that only smell opens up the realm of space for us. The total "core and foundation of space" involves more than smell, constituting in fact "the total body, the brain, gestures, and so forth." Smell, however, is the most fundamental of our bodily foundations of space, by virtue of the fact that "smells are not decodable." This does not mean that smells have an indecipherable code, but that they could never have been coded in the first place. We may not want to adopt such an extreme conclusion, but say rather that smells are one of the least codable forms of sensation, both because smells are difficult to manipulate, and because our sense of smell is poorly developed. It is no accident that we use audible sounds and visible signs for communication. Our senses of hearing and sight are able to make fine distinctions and we are easily able to manipulate sounds and sights.

I can make a remarkable number of different sounds with my mouth, discern them with my ears, and even call them up at will in my imagination without making any exterior noise. But I can produce very few smells, and once I have ceased to smell them I can barely imagine them at all. Then, too, when I am smelling them I have few universal terms with which to describe them. I am more likely to give the smell the name of the thing that produces it: it is a "pine scent" or a "rose scent." Because it cannot be represented in the imagination, and because of its close alliance with the thing that produces it, the sense of smell is perhaps, of all the senses, the most closely linked to direct experience. Since rhetoric uses codes to shift the relation between sensation (sign) and thing (signified), we may agree with Lefebvre who states, "anyone who is wont (and every child falls immediately into this category) to identify places, people and things by their smells is unlikely to become very susceptible to rhetoric."[98]

The close identification of smell with its source does not insulate it from a danger common to all the senses. When we value the sensation as distinct from the thing that produces it, a gap opens up between the sensation and the thing. This gap can then be manipulated just like the gap between sign and signified in language. The manipulation is easiest in the case of industrially produced sensations, which have no previous source in a thing. Perfumes, for instance, are linked to attractive "signifieds"—"woman, freshness, nature, glamour, and so forth"—through words and images. Because Lefebvre is confident that smells themselves cannot be coded, he warns us that this is an exterior discourse about the smell and not the smell itself: "a perfume either induces or fails to induce an erotic mood—it does not carry on a discourse about it." Since discourse about industrially produced smells is easier to control than discourse about natural smells, which already have a well-known source, we find in industrial societies an "immense deodorizing campaign, which makes use of every means to combat natural smells." The unmanipulated smell of a sheet dried on a clothesline in a rural area is replaced by the "fresh" scent of a dryer sheet.

Lefebvre occasionally overstates the importance of sensation—going so far as to treat the body and sensation as identical—but he usually maintains a balance between the senses and the human body as forms of resistance to coding. If the sacred place is to resist the twin temptations of dematerialization (by overlooking the role of the body in establishing it) and manipulation (by reducing it to an image to be read), then it must at all costs resist the coding inevitable in spaces of the modern world. Given that coding deemphasizes the body, one might expect that any incorporation of the sacred within a linguistic context would easily degrade the role of the body in its rites. But the Neoplatonists, the first thinkers to incorporate the sacred place into their philosophical systems, save the bodily character of the sacred place by their very incorporation of it into a linguistic context, the web of symbols established by their mythic tradition. These symbols can successfully reveal the divine only if they are overlaid on the properly material context of natural forms, which come about in place. The natural forms remain thoroughly material, but in the context of the sacred, they indirectly reveal the divine. Natural forms only dematerialize if they attempt to reveal the immaterial gods directly, without the mediation of symbols. Likewise, the symbols only become mere tools of power to the degree that they overcome and supplant the system of natural forms. The story of the Neoplatonists is, in large measure, the story of this tension between natural form and symbol, between place and the sacred, and the dissolution of this tension on the threshold of modernity.

CHAPTER 2

THE NEOPLATONIC BACKGROUND

The Platonic identification of unity and goodness means that unity of some sort will be a goal in any Platonic philosophical system. But not everything is capable of the same degree of unity. The Neoplatonists are traditionally distinguished from other Platonists by their claim that perfect unity must be beyond being, since all being involves some sort of multiplicity. They give the personal name of "One" to this perfect unity which, since it is prior to all multiplicity, must be regarded in some sense as the cause of all multiplicity. Between the unity of the One and the multiplicity of our world lies the divine intellect, the "One-and-Many." In the divine intellect, the forms of things are distinct from one another, but each form has all the other forms within it. The form of an oak tree, for instance, is distinct from the form of a rose bush, but it has the form of the rose bush within it, and vice versa. If we could think at the level of the divine intellect, we would remain ourselves, but in ourselves we would be one with all other things and with the gods. Presently, however, our thought is conditioned by the bodies we inhabit, and bodies can only exist in places, which ensure one kind of unity while prohibiting another. Place holds together in a unity whatever it contains, but it isolates its contents from all other things. A garden, as a place, holds together all the plants that it contains, but it keeps those plants from being one with anything outside the garden. The Neoplatonists follow a longstanding tradition of exploring the nature of this bodily place, whether this is the only kind of place, and most importantly for our purposes, whether this kind of place must be overcome if we are to achieve unity with each other, with all things, and with the gods.

Place

Plotinus (204–270) wrote no treatise on place that we know of. Perhaps for this reason most scholarly accounts of place in the Neoplatonists spend little time on Plotinus.[1] His philosophical system, however, is rooted in an understanding of place, and we may look at his brief explanation of it as a prelude to the lengthier accounts of it later in the Neoplatonic tradition. In one of his longest and most central treatises, he briefly takes up the question, prompted by Plato's *Parmenides*, of what causes one thing to be in another.[2] He tells us that everything that is generated, whether by human artifice—as in the case of a chair—or by nature—as in the case of an oak tree—must be in something else. Either the product is in its cause, or it is in something other than its cause. In the case of the oak tree, it must be either in the oak tree that produced it, which it is not, or it must be in something else, in this case the earth and air that surround and sustain it. Why must it, at all costs, be in something? Plotinus explains that, "because it comes about by another's agency, and requires that other for its coming about, it also requires that other at every point."[3] That is, everything that comes into being is in something else out of need. The oak tree needs the earth to sustain it; otherwise, it literally falls. To be in something different than the thing that produced you: this is how visible things are in place, though Plotinus never puts it quite so explicitly. All the things that we see are in place because in each case the product is in something different from the producer. There is one exception: the visible cosmos as a whole cannot be in another visible thing, since it is by definition the sum of all visible things. If it is in anything, then, it must be in its cause, which is, as Plotinus understands it, its invisible soul. Plotinus does not avoid the language of place when he introduces such invisible beings, and he does not suggest that his use of "place" in these cases is less literal than when describing bodily places.[4] The cosmos is in its soul, that soul is in the universal intellect, and the intellect is in the One. The One presents a unique case. It cannot be in its cause, since it has no cause, and yet neither can it be in anything other than its cause, since other things do not exist until it causes them. As a result, it possesses special properties relative to place. It is the place of all things, since it is the place of the intellect, the intellect is the place of the soul, and so on, down through all things. By being the place of all things, the One is also everywhere. For "if it were prevented, it would be defined by another, and. . .God would no longer be in control, but would serve those who come after him."[5] The independence of the One requires that it be everywhere, not as though it were in any particular place, but "as not being absent from anything."

Plotinus, then, lays the ground for three kinds of place: (1) when something is in something different from its cause, which is the case for all visible things except the cosmos itself; (2) when something is in its cause, which is the case for the cosmos, the soul, and the intellect; and (3) when something is in all things, though not in such a way as to be dependent on them, which is the case solely for the One. Everything, then, including the One, has some involvement with place. It is left to the later Neoplatonists to put these kinds of places into the more technical terminology given to the topic by Plato and Aristotle. Some of them follow Plato in identifying place with extension, and so give place some of the characteristics of space. Others blend Aristotelian language with the un-Aristotelian identification of place with form, and so bind place closely to the nature of the thing it contains. Whichever their approach, every major Neoplatonist after Plotinus devotes at least a few lines to the treatment of place. Some of their commentaries on Aristotle's *Categories* survive, but for the most part, their views survive only because the sixth-century Neoplatonist Simplicius summarizes them in his commentary on Aristotle's *Physics*, from which I have drawn most of the material for the following accounts.

Absolute Place

Contemporary studies of place in Neoplatonism have tended to focus on its role in laying the groundwork for the modern concept of space as an immobile passive extension. Shmuel Sambursky's *The Concept of Place in Late Neoplatonism* pushes the Neoplatonic philosophers as close as possible to modern theorists of space like Isaac Newton, sometimes misreading them in the process. Edward Casey's *The Fate of Place* avoids the excesses of Sambursky, but he does set his discussion of the Neoplatonists in the context of the development of space, entitling his chapter on them: "The Emergence of Space in Hellenistic and Neoplatonic Thought."[6] When Sambursky and Casey speak of space in Neoplatonism, they mean the place of the cosmos, which is motionless because it has nothing against which its motion could be compared.

If Plotinus were asked the place of the cosmos, he would undoubtedly reply: "the soul." He does not explain in detail how this place differs from the place of bodies within the cosmos, and for this reason he leaves unanswered the difficult question of how an immaterial being can be the place of a material being. Aristotle avoids the question by limiting place to the surrounding of one body by another, and as a result he denies that the cosmos can be in a place at all.[7] The later Neoplatonists Syrianus and his student Proclus (412–485) solve the problems associated with identifying soul and place by claiming that the place of the cosmos is an active extension, which lies *between* the soul and cosmos. Since we find a more developed understanding of this theory in Proclus than in his teacher, I will rely on his explanation of it in what follows.

Simplicius gives us the Proclan theory of absolute place in his commentary on Aristotle's *Physics*.[8] Proclus argues first that there must be an extension which traverses the entire cosmos, because otherwise there could be no motion from one part of the cosmos to another. Motion requires something to measure that motion against. If there is no cosmic extension, then there is nothing against which to measure motions of the largest order, the motions of the four great regions of the cosmos: fire, earth, air, and water. This extension, which stretches the length of the cosmos, must be a body if motion is to be measured against it, but it is no ordinary body. It must be immobile (because if it were in motion it would itself require a place against which to measure that motion), indivisible (because if it were divided there would be a dividing body that would itself need a place), and immaterial (because material bodies are divisible).[9]

What could such a body be made of? Not earth, since earth is the most material of all the elements, and the cosmic place "is the most immaterial of all bodies, both those that are moved and those that are immaterial among those that are moved."[10] These latter bodies are the heavenly bodies—the sun, moon, planets, and stars—which move, but are not properly material since they undergo no change other than circular motion. They are made of fire, an element Proclus looks to as a clue to uncovering the material constitution of the cosmic place: "if light is the simplest of the elements, since fire is the most bodiless of the other elements, and light is more bodiless than fire itself, it is clear that place is light, the purest among bodies."[11] The visible cosmos, then, is permeated with a motionless, indivisible, immaterial form of light.

Without the cosmic place, the soul of the cosmos could not set anything in motion. The motionlessness of place is not passive but active: it causes other things to move. And so place is alive, "the primary participant of life, giving motion to its parts, making each to reveal its proper motion in place."[12] The parts, in turn, by their motion, attempt to become the soul by becoming like the cosmic place. Each part wants to be "in the whole of it." The whole will always elude the efforts of the parts to become it because, "on account of the natural property of extension, it is not possible for a part to come to be in the whole." The cosmic place, then, is independent of the bodies in the cosmos as well as its soul, even as it mediates between them by its likeness to them. By its motionlessness it resembles soul, but by its corporeality it resembles the cosmic body. This place is essentially sacred—it irrupts into our ordinary divisible places when the universal intellect manifests itself as light. It is thus able to explain unusual events like the visible illumination of Plotinus' face when he spoke, as reported by his student Porphyry: "in speaking, there was a manifestation of his intellect, which shined a light on his face."[13] It can also explain the practice, described at length by Iamblichus, of manipulating visible light so as to see the will of the universal intellect.[14]

This cosmic place, which we may call absolute because of its motionlessness and presence in every part of the cosmos, is the primary interest of Sambursky and Casey in the Neoplatonists. The Newtonian concept of absolute space more nearly resembles the passive extension or void that had appeared in Western philosophy before the time of Socrates than it does the Neoplatonic cosmic place, but the latter does demonstrate the continuing presence in Western thought of a motionless place against which all motion may be measured. Most of the Neoplatonists do not mention this kind of place, and those that do hold it together with a different, Aristotelian concept of place, which they employ in discussing the place of particular things. Even Proclus adds to the place of the cosmos the "primary place of each thing," which is the "extension between the limits of the surrounding thing."[15] Sambursky translates "each thing" (ἑκάστου) as "everything," which gives the misleading impression that Proclus is still speaking about the place of the cosmos.[16] Sambursky's interest in absolute space plays him false here, as it does in the introduction to his section on Proclus, in which he treats the definition of a particular place as no different from the definition of absolute place. More misleading is Sambursky's description of an absolute place theory in the passages from Syrianus quoted by Simplicius in his commentary. Syrianus does indeed possess a theory of absolute place, the one his student Proclus adopted and expanded, but we find it only in his commentary on Aristotle's *Metaphysics*. When we turn to the thumbnail sketch of Syrianus provided by Simplicius in his commentary, we hear only of a nested series of Aristotelian-style places within the cosmos, places that are by no means absolute. Sambursky's reader will find himself misled if he attends to Sambursky's notes and not the text itself. What we find in Syrianus, as Simplicius presents his work, is a theory of place which has a great deal in common with the distinctive Neoplatonic development of Aristotle, and which has nothing at all to do with the notion of absolute space.

Syrianus distinguishes two kinds of place: the "primary and proper place," and "place as width (ἐν πλάτει τόπος)."[17] When we speak of place as width, we mean simply the region or landscape a thing happens to be in. When we say that a ship is in the sea, for instance, we do not necessarily mean that it is in a specific part of the sea. The sea is simply a width, which generally encompasses the ship. If, on the other hand, we want to identify the exact shape taken up by the ship within the larger place of the sea, we seek its "primary and proper place." Simplicius amplifies Syrianus' account here with his own commentary, adding two important details. When most people speak of place, they mean place as width. They think of the earth as in the cosmos, but they do not think of the earth's proper place, "the place which properly positions its parts."[18] The proper place of a thing here does more than the Aristotelian job of surrounding it. It positions the parts of the thing so that it can exist in the best way possible. Place here combines elements of the "limit of the surrounding body" theory and a new sense of place as defining the thing from the inside. Since most people speak of place only as width, they overlook both these features of place, and think that place is "separate from what is in place." Only place as width can be separated from the thing that is in it. A thing cannot be separated from its definition, which is its very form. Likewise, it cannot be separated from its own shape, because it *occupies* that shape.[19] The second detail added by Simplicius is the immobility of place as width. Place as width allows the things within it to move only because it itself remains motionless, as something against which motion can be measured. If the sea were in motion from one part of the globe to another, we could not tell if the motion of the ship were its own motion or that of the sea. The primary and proper place of the ship, however, moves together with it.

Simplicius shows the relationship between these two kinds of place by setting them up in a relation between three different bodies. The sea (body #1) in which the ship (body #2) sails is separable from the ship, but it is inseparable from its own proper place, the basin of earth (body #3) which it fills. Simplicius explains: "as Syrianus also said, place as width [body #1], according to which even motion in place comes about, is proper to another more universal body [body #3], which is both inseparable from it and is moved together with it, if it should happen to be moved, but is separate from the body [body #2] which is *in* that body [body #1] and is unmoved in relation to it."[20] That is, the sea is separate from and immobile relative to the ship contained in the sea. Sambursky, in his eagerness to chart the development of absolute space, defines the "more universal body" not as what in this example is the basin of earth, but as "absolute space."[21] He goes on to misunderstand the relation between "place as width" and the "more universal body." He suggests that the more universal body (in this case, the basin of earth) must extend further than place as width (in this case, the sea), when in fact the place as width has the more universal body as its own proper place, and must be coterminous with it.[22]

Sambursky's introduction to his section on Syrianus gives us the context in which such misunderstandings can come about. He tells us that the proper place of each thing—what he calls "place in the restricted sense"—moves along with the thing, while place as width—what he calls "place in the broader sense"—"is immobile and must be understood as the space encompassing a body in motion."[23] By failing to see that place as width is immobile only relative to the object in it, Sambursky takes Syrianus to be saying that place as width is absolutely immobile and belongs to the Neoplatonic theory of absolute place. Syrianus does indeed have a theory of absolute place, but it is to be found only in his commentary on Aristotle's *Metaphysics*, and even there it is an active extension and not the passive extension of Democritus and, later, of John Philoponus.[24] To misread his theory here as describing absolute place is to overlook the importance of the more traditional Neoplatonic theory of nested places: the ship in the sea, the sea in the basin, the basin in the earth, and so on. Because he misreads Syrianus' theory of place as width, Sambursky draws a startling and erroneous conclusion: "this train of thought is a clear transition from the concept of place to that of absolute space."[25]

Place as Definition

The notion of absolute space ripples underneath the surface of discussions of place within the Neoplatonic schools, but it seldom surfaces, and its importance for the modern period should not lead us to overstate its importance for the Neoplatonists. More important for the Neoplatonists, and more overt, is the careful working out of a synthesis and development of Platonic and Aristotelian accounts of place. We find such syntheses in the work of Iamblichus, Syrianus, Proclus, and Damascius, all gathered together in Simplicius' commentary on Aristotle's *Physics*. Iamblichus provides one of the broadest of these accounts, addressing all three of Plotinus' forms of place. Since Iamblichus also provides the major Neoplatonist account of the sacred place, I will give his account special prominence. After examining Iamblichus, we will turn to Damascius, who identifies place with natural place, and dramatically expands the character of natural places to include the proper placement of parts within their whole. Place then becomes not what allows a body to exist, but what allows it to reveal its true nature.

Iamblichus (active *ca.* 300) begins with Plotinus' first kind of place: the sort of place we encounter in our bodily experience. We say that Socrates is in the house, in the bath, or in the theater, and we mean that one body (Socrates) is in another body (the house, bath, or theater).[26] An interaction with bodies outside of place is impossible. Iamblichus says: "every body, insofar as it subsists as body, is in place."[27] Because there can be no body without place, Iamblichus ranks place not with bodies, but with the cause (ἀρχή) of bodies. This step takes him well beyond the Peripatetic view of place as the exterior surface of bodies—their container—ranked with them and exercising no power over them. Iamblichus does not spare his criticism of such a view: "those who do not make place akin to the cause, but drag it down to the limits of surfaces or empty containers or also extensions of any kind introduce foreign opinions and miss the whole point of the *Timaeus*."[28] Iamblichus may seem here to be criticizing naive readers of Plato's *Timaeus*, but what is at stake is more than a misreading of Plato. To suggest that place is a limit merely in the sense of a container, which surrounds a body like a piece of clothing, saps place of any connection to the nature of the body concerned.[29] By supplementing the Peripatetic account of place as container, Iamblichus comes up with a richer definition of place, one that unites it to the nature of the body concerned. He defines place as "a power, bodily in form (σωματοειδῆ), which holds up and props up bodies, raising up the falling and gathering the scattered, filling them and surrounding them on every side."[30] This definition incorporates the Peripatetic view of place in only two of its many elements: the "bodily in form," and the "surrounding them on every side." The rest of the definition describes the new, active role of place Iamblichus suggested in his introductory comments. Without place, bodies cannot be held together in their existence. They need place to "hold them up." In a strange echo of a biblical commonplace—"he will gather the scattered ones of Judah"[31]—Iamblichus speaks of place "raising up the falling and gathering the scattered." On the basis of this echo, Shmuel Sambursky says that Iamblichus incorporates Jewish sources in his discussion of place.[32] The phrase is suggestive, particularly because of the absence of any reference to "gathering" in the Platonic and Aristotelian accounts of place, but in the absence of a historical investigation into Iamblichus' ties to Judaic Platonism and its texts the implications must remain tantalizing but uncertain.

Iamblichus does not explain what he means by "holds up," "props up," "raising up," and "gathering," and Simplicius provides no commentary. We may conjecture from Iamblichus' introductory comments that the "holding," "propping," "raising," and "gathering" refer to the act of sustaining the very existence of bodies. Just as solid ground holds up, props up, and raises a body so that it does not sink into the earth, so it is with place. Whereas solid ground simply acts on a body from without to keep it from sinking, place supports a body on all sides and from within. It gathers the different parts of the body together so that it does not disperse into the indefinite.

What begins as a seemingly familiar description of Plotinus' first kind of place, then, ends in the less familiar grounding of place within the incorporeal realm of definition. Iamblichus' contemplation of Plotinus' second kind of place, referred to by Simplicius as an "intellectual contemplation" of place, remains in the incorporeal realm. Iamblichus extends the term "place" to things that are not bodies on the basis of a reflection on the term "surrounding" (περιέχων). The Peripatetics use the term to describe what the container does to the contained, and Iamblichus has included this in the role of bodily place, as we have seen earlier. But the term "surrounding" can be applied to more than bodies. In the broadest sense, anything that has definition is surrounded by what defines it: "we must extend the whole nature of place, not metaphorically (ὁμωνύμως), but by the same definition of the genus, to all beings whatsoever that exist as one being in another. For the relation of the surrounded to the surrounding is one, and it is everywhere the same, though it varies according to the different subsistences of its participants."[33] An oak tree's place differs from a triangle's place only in as much as an oak tree differs from a triangle: the oak tree has bodily extension, while the triangle does not. The place of the oak tree will then surround and sustain it in a bodily fashion, while the place of the triangle will not. The relation between the oak tree and its place, and the triangle and its place—the relation of surrounding, in other words—remains exactly the same. Iamblichus rules out a merely metaphorical relation between the two kinds of place, emphatically declaring that he means exactly the same thing by "place" when speaking of bodiless beings as he does when speaking of bodies.

Iamblichus does not mention Plotinus' third kind of place, appropriate only to the One, beyond the places of visible and intelligible things, but he does call the One the highest instance of the second kind of place: "we should consider, in an even higher sense than this, that the substance of place is as the substance of God, since it is one in form and holds together all things in itself, and limits all things by one measure."[34] The unity of the highest holds all other things together, and so does the job of a place. Iamblichus calls it a "divine place," which is "cause of itself, comprehends itself, and subsists in itself, and is no longer inseparable, but has a subsistence separate from beings."[35] This divine place is the first kind of place that is separable from the form of the beings in it. The other two kinds are inseparable from their contents, and so do not merely locate them, but are responsible for their existence in the first place.

Place as Measure of Position

Damascius, the last head of the Neoplatonic school at Athens before it closed to the Hellenic philosophers in 529, begins his treatment of place with the apparently Aristotelian claim that it is the measure of bodily extension. Damascius denies the Iamblichan claim that there can be places of non-visible things. In the course of restricting place to things having the kind of extension that makes them visible, Damascius carefully distinguishes the extension that belongs to bodies in place from other forms of extension. As a broad category, extension is applicable to every body that comes to be and passes away in the visible world. We may divide this broad category of extension into two subcategories based on the familiar distinction between "substance" (*οὐσία*) and "activity" (*ἐνέργεια*). Extension in activity is extension in time, and breaks down into its own two subdivisions according to whether it is continuous or occasional. For instance, living is a continuous activity, while going for a walk is only an occasional activity performed from time to time. Since these forms of extension are limited by time and not place, we may pass over them here.

Extension in substance, on the other hand, involves place in one of its subdivisions. We may divide extension in substance into two subheadings: extension in number and extension in mass. Extension in number occurs any time we distinguish a number of things without referring to any qualitative differences between them. When I distinguish five apples, for instance, I use number to limit the extension of the apples in their multiplicity. Extension in number is not sufficient for place, since "numbers, though divided (*διακριθέντες*), do not seem to have position, since they are not extended [in mass] and set apart, save when they too take on size and extension [in mass]."[36] When I conceive nine cubes, for example, I distinguish nine different cubes, but these cubes may not have a given size, and they do not possess mass. If I shape nine cubes out of clay, on the other hand, these cubes will have a certain size and mass, and so will take up positions.

What is this extension in mass, which constitutes the essential condition for taking up a position in space? Damascius explains: "extension by size and mass comes at once to be in a position owing to the spreading (*διάρριψις*) of its parts one from another."[37] Extension in mass seems to differ in two ways from extension in number: (1) extension in mass involves "spreading" (*διάρριψις*), while extension in number involves "division" (*διάκρισις*); (2) extension in mass involves parts, while extension in number involves wholes. When I conceive nine cubes, I divide nine wholes from each other without necessarily considering their parts at all. When I shape a cube from clay, however, I attend to each part of that cube with my hands, spreading the clay until it becomes each part of the cube. This extension in mass itself may finally be subdivided into two subordinate forms, one which lasts as long as the thing itself does, and one which belongs only occasionally to the thing. As an example of the first form, Damascius mentions the relative position of the parts of the body. My head is at the top of my body, my feet are at the bottom, my liver is on the right, and my heart is in the middle. This "directional" extension of my body lasts as long as I do: my liver will never replace my feet at the bottom of my body. The second form of extension in mass describes the category of "where" from Aristotle's *Categories*: being at one time in the house, and at another in the marketplace.

It is the first form of extension in mass, though, that turns out to be more relevant to the essence of place. This extension is not homogeneous, since it is the extension belonging to a whole that has differentiated parts within it. In order to describe place as the limit to such an extension, Damascius borrows from the Aristotelian Theophrastus a conception of place as what orients parts toward each other and toward their whole.[38] Having said that certain forms of extension in mass belong to the nature of a thing, Damascius distinguishes several forms of "lying somewhere" (κεῖσθαι), some of which involve belonging in a specific place. Anything whose parts "are set apart and extended from each other" may be said to "lie somewhere." This includes all forms of extension in mass, which transform them from being in themselves (as intellectual forms) to being in something else (whatever supports and sustains their bodies).[39] Thus, they lie in something else and so are in Plotinus' first kind of place. Where they lie is their "position" (θέσις). So far, Damascius has not described anything more than geometrical space, in which a body may conceivably take up a position at any point. Damascius takes up the term "place" as a narrower category within the larger category of position. To have "place," a thing must first lie somewhere and have position, but place additionally "defines, measures, and ranks position."[40] Position specifies nothing about the nature of the thing concerned, while place does nothing but specify the nature. It does not cause the thing to exist in the first place. A thing may exist while possessing position alone: "we say that each thing lies somewhere even if it lies anywhere without order, but each thing is said to have its own proper position (εὐθετισμός) when it takes its proper place."[41] Damascius again refers to the parts of the body to clarify his point. He says: "owing to place, then, which properly positions each of my parts, my head comes to be at the top of my body, and my feet below." Place is not simply the "where" of my body's parts. It is the power that actively positions them where they, by nature, ought to be.

There can be no place without wholes and parts.[42] The whole provides a dimensional structure within which the parts can find their proper places. This holds for organic wholes within the cosmos, like the human body—within which we find a proper position for the liver, the heart, the feet, and the head—as well as for the organic whole that is the cosmos itself. As Damascius puts it: "the parts of the cosmos, through place, hold their own proper position within the whole."[43] Each of the four regions within the cosmos has a proper position relative to the whole. The air belongs close to the circumference (or "up"); the earth belongs at the center (or "down"). These regions may then serve as wholes, with their own proper relation to their parts. In the region of the earth, for example, some beings—like polar bears—belong in the north, while others belong in the south. Because place has this power of organizing the parts within the whole, Damascius says that it supplies " 'falling together' (σύμπτωσις), that is, 'placement together' (συντόπωσις)."[44] Place gathers together a group of otherwise disparate beings into a whole. For the first time among the Neoplatonists, we see place functioning not as the cause of a single being (as in Iamblichus), or as the homogenous region of a single being (as in Syrianus), but as the gathering force which draws several different beings into an organized structure: a whole.

Damascius refers only to organic wholes that come about independently of human building, whether they be particular bodies—like the human body—or the body of the entire cosmos. It is not clear whether he intends his model to be applicable to organic wholes constructed by human beings. That is, it is not clear that we may refer to a house or a town as a place. We can easily think of the house analogously to an organic body: the stove should be near the refrigerator and the food cabinets, the towel rack should be near the bathtub, and so on. Likewise, we can think of a city as requiring residences to be removed from heavy industry, but near retail stores. Damascius, however, never says this explicitly. We find a similar situation in Plotinus, who refers to human constructions occasionally, but suggests that they are properly discussed within the context of the organic whole that is human nature.[45] Like his peers and predecessors, Damascius seems to have passed over the discussion of human constructions as a subordinate matter, perhaps easily deducible from his conclusions concerning independent organic wholes. His chief contribution to the Neoplatonic theory of place lies, rather, in his more general discussion of place as the gathering of parts in their proper relation to each other and to the whole.

After recounting the Iamblichan theory of place in his commentary, Simplicius adopts the Damascian theory of place, and rejects Iamblichus. Simplicius flatly denies the Iamblichan claim that beings in the intelligible realm are literally in place. He seeks "a more commensurate use and nature of place—speaking of it properly and not metaphorically—and we say that everything that does not come to be and is partless has a united substance and state of being, is in itself wholly and at once, and for this reason needs neither place nor time."[46] He follows Damascius, for whom place has no purpose other than the ordering of parts to whole. The partless does not need to be ordered, since the very term "order" makes no sense without a number of parts to be set in order. Since the term "place" refers to an ordering principle, it too makes no sense when applied to partless beings. At stake here is the distinction between position and place. Since Iamblichus regards place not merely as an ordering principle, but as responsible for the very existence of a thing, he cannot meaningfully claim that things can have positions without places. If a thing ceases to have a place, it ceases to exist. Damascius' more complex treatment of place suggests that things can have existence without having place. This placeless existence is existence in position alone. The Damascian scheme leaves open the possibility that we could break down the perfection of things by moving them out of their proper places: putting the polar bear in a zoo near the equator, for example. Our understanding of a thing's place, in this sense, requires that we know the meaning of the thing in relation to the whole context of meanings which finds its embodiment in the cosmos. Conceived in this way, place plays a complementary role to the sacred place—a certain region, city, or temple that visibly grounds the context through which the existence of things can be perfected.

The Sacred Place

While the Neoplatonists use the phrase "sacred place," they do not make it a theme of philosophical analysis as they do in the case of "place" *tout court*. The analysis of place has a long history in the ancient schools, receiving at least some treatment in the Platonists, Peripatetics, Stoics, and Epicureans. The sacred place, on the other hand, is generally left up to the traditional religious authorities. In Plato's *Republic*, Socrates tables any discussion of laws "having to do with the establishment of temples" with the words: "we have no knowledge of these things, and. . .we won't be persuaded to trust them to anyone other than the ancestral guide."[47] Only when the later Neoplatonists include an account of religious practice in their study of reality does the sacred place begin to crop up more explicitly. It never emerges as a theme in its own right, but it implicitly sets itself in tension with their analysis of place.

The two terms "sacred"(ἱερός) and "place" (τόπος or χῶρος) put together concepts which are, in Neoplatonic terms, virtually contradictory. The term "sacred" never describes the gods themselves—we never speak of "sacred gods"—but certain visible things that have an unusually explicit tie to the gods. Every place, on the other hand, is tied to the gods, since the term "place" pertains to the proper manifestation of form, and form is always either divine itself (in the universal intellect), or an image of a divine form (in the soul or a visible thing). Proclus has left us with perhaps the most coherent and complete account of form, and I will summarize it here in the course of distinguishing form from the sacred.[48] An independently existing intelligible form is what first gives existence to a visible thing, which is in turn an image of the form combined with some material. Form not only gives the thing its existence, but also provides its specific characteristics.

A rose bush, for example, owes both its being and its being a rose bush to form. Now some rose bushes are prettier than others, and some are healthier than others. An unhealthy rose bush has not ceased to be a rose bush, nor has it ceased to exist, so the form also seems to be responsible for health and beauty in the rose bush. On the basis of these three principles—specific characteristics, health and beauty, existence—Proclus distinguishes three kinds of form.

The "most specific" forms are "whatever can be participated by individual forms, like 'man', 'dog', and things of this sort. The production of these immediately begets monads in individuals—the 'man' in particular men, the 'dog' in the many dogs, and the 'horse', and each of the others as well."[49] These are the forms that are immediately participated by individual existing things. The term "individual" (ἄτομος) in Greek literally means what "cannot be cut," meaning that these forms cannot be subdivided into further forms before they are participated by existing things. The form of a dog cannot be divided into into forms of "ear," "tongue," and "tail." The form gives existence to the dog, but the tail does not need its own form in order to receive existence: its existence is given to it through the form of the dog. Proclus does not deny that the tail has form in a restricted sense, since our souls can consider it and its purpose, but this form appears only in our souls, not the divine intellect.[50]

Certain qualities in things also have forms. There are two kinds of qualities, since "some qualities, such as likeness, beauty, health, and virtue, complete and perfect their substances; others, like whiteness, blackness, and the like, exist in substances but do not complete or perfect them."[51] Qualities that do not complete or perfect their substances do not have forms, for they, like the dog's tail, only exist because the thing they qualify already exists. Whether I am white or black, for example, does not determine my existence as a human being; one is not the imperfection of the other. Other qualities, like "beauty, health, and virtue," do perfect their substances. For a human being not just to exist, but to exist in the best way possible, it must be beautiful, healthy, and virtuous. These three qualities are Proclus' "intermediate" forms, which "have wider application than" the specific forms, "but are not active in all things." Health, beauty, and virtue are of course not active in those things that exist without being healthy, beautiful or virtuous. The forms that are active in all things, and so are the most general forms, are those that are convertible with existence itself. Proclus gives three: "being, identity, and otherness." Plotinus follows Plato in listing five: being, identity, otherness, motion, and rest. Every being is different or "other" than all other things, but is identical with itself. When we say that the being remains the same as itself, we introduce the new form of "remaining" or "rest." It does not move away from its own form. The "rest" of the being does not mean that it is simply inert. By existing it has the activity or "motion" of sustaining itself in its existence. Plotinus' early work suggests that the form of "being" may be prior to the other four, but his later work treats them, as Proclus does, as interchangeable names for the divine.[52] All of these forms, from the most specific to the most general, are divine in that they cause the things we see without sharing either the motion or the multiplicity of the things we see.

To the degree that form becomes mingled with matter, it produces a new entity which is not divine, though it participates in the divine by participating in form. Should this participation render it sacred? If it did, virtually everything we encounter would be sacred. All trees, as images of the divine form of tree, would be sacred. All people, as images of the divine form of human being, would be sacred. The earth below and the sky above would be sacred, and even streets and cars could be sacred in a secondary sense as the products of sacred human beings. Were the Neoplatonists to use the term "sacred" in this sense, it would have little in common with the traditional Hellenic term for the sacred place—the *temenos*—whose name derives from the verb "to cut off" (τέμνω). The Hellenic sacred place is "cut off" from ordinary concerns. If we use the term "sacred" simply to refer to all organic wholes, we reduce the term to a synonym for "image" and render the traditional sacred place—the temple precinct or *temenos*—meaningless.

The Neoplatonists do not take this road. While all organic wholes are images of the divine form, certain things share a more arcane relation to the divine form expressed through a different vocabulary. Terms like "symbol" (σύμβολος or σύνθημα), and "sign" (σημεῖον), more or less synonymous with each other, describe a relation to the divine not evident in the image. The term "symbol," in both its Greek forms, *symbolos* and *synthema*, fertilize the Neoplatonic imagination because of its use in the Chaldean Oracles, a second or third century text which becomes a sacred scripture for the Neoplatonists.[53] The Chaldean Oracles treat symbols as identical with forms, but the symbols of the late Neoplatonists "*perfect* the cosmos rather than simply *enform* it," as Andrew Smith has noted.[54] Iamblichus explicitly distinguishes the two when discussing the arts of prophecy: the examination of animal entrails, bird behavior, and the motions of the heavens. In these phenomena, the gods manifest themselves, but not in the way that they manifest themselves by creating the thing in the first place. As Iamblichus puts it, the gods "generate all things through images, and they likewise signify all things through symbols."[55] The eagle we see perched on the branch of a dead tree is a composite of material and an image of the divine form of "eagle" in the universal intellect. The image reveals itself to us through the visible shape of the eagle we see in the forest. When that eagle takes off from the branch and flies to the left, it becomes a sign, an additional manifestation of the gods beyond its mere character as an eagle. This additional manifestation can "symbolically manifest the will of God and the foreshadowing of the future." The sign is not self-evident. Iamblichus reminds us of Heraclitus' dictum, that the gods "neither speak, nor conceal, but signify."[56] Speaking is the work of an image. No occult process of interpretation is needed to recognize an eagle sitting on the branch of a dead tree. We may need binoculars to get a better look, but we never need to go beyond our senses and the meaning they impart to the soul. The sign, on the other hand, is manifest to the senses, but it needs interpretation to connect it with its meaning, which resides in the intelligible realm of the gods. It does not speak, then, but signifies. Because the sign has no direct effect on the existence of either the thing that exhibits the sign or the human being that interprets the sign, we cannot say that it gives existence as the forms do.[57] It shapes the existence of the human being for the better, playing a perfective rather than a formative role.

Plotinus: Light and Stillness

Not every Neoplatonist requires this bifurcation of the world into images and symbols. Plotinus, in particular, uncovers characteristics of the visible world that directly reveal the positive presence of the divine intellect without a superimposition of symbolic content. Beauty is the most prominent of these, though not as understood by Plotinus' contemporaries. Beauty "is said by all, so to speak, to be a proportion (συμμετρία) of the parts to each other and to the whole, supplemented by good color."[58] This common definition of beauty catches a glimpse of the truth with its mention of "good color," but by making proportion the chief criterion of the beautiful, it mistakes a subordinate manifestation of beauty for its cause. The limited explanatory power of this definition reveals that it has not grasped what is essential to beauty. If beauty is restricted to things that have proportion, then "nothing simple, but only the composite will be beautiful." The beautiful will then exclude "beautiful colors," "the light of the sun," "gold," "lightning," and "stars," all of which "are simple and do not have their beauty from proportion."

Proportion is a characteristic of bodies *as* bodies, and of souls in so far as they are concerned with bodies. It has no place in the properly incorporeal realm of the intelligible. When commenting on Plato's claim that the person closest to leading a bodiless life is "a man who will become a lover of wisdom or of beauty, or who will be an artist and prone to erotic love,"[59] Plotinus identifies the artist as a musician, in love with audible proportion, but distinguishes him from the lover of wisdom, who can see directly into the intelligible realm. The musician, Plotinus says, is "always fleeing the inharmonious and what is not one in songs and rhythms, and pursues good rhythm and the well-shaped."[60] The harmony sought by the musician is self-directed. Unlike a spinning whirlpool, it holds its various parts in tension without transporting them into another realm. Someone else must prod the musician to "separate the matter in those things whose proportions and structures lead to the beautiful which is in them," using other means than the beauty of harmony. Plotinus' understanding of harmony here relies on Plato's discussion in his *Republic*, in which harmony serves as political virtue and not an aid to salvation. Education in music has pride of place in the ideal republic, because "rhythm and harmony permeate the inner part of the soul more than anything else."[61] They do not however, lead the soul to the intelligible. When Socrates takes up the question of how we will educate students to ascend to the intelligible, he must adopt an entirely new system of education. Plotinus himself, when he briefly treats political virtues, characterizes them as confined to an order *within* the visible world, and different from the higher virtues, which lead the human being toward the divine.[62] Their defining characteristic is that they give order and measure—proportion, that is—to our lives. Proportion is not a characteristic of the intellect, but belongs properly to the visible realm.

The common definition of beauty is more successful when it mentions the need for "good color," but even here it falls short, since the beauty of colors does not arise from the colors themselves. Their beauty comes "from shape, and from the overpowering of the darkness in matter by the presence of light, which is bodiless, structure, and form."[63] The apparent beauty of a color is sometimes owing only to the beautiful shape of the colored object, but when colors are beautiful by themselves, their beauty comes from light. For the ancients, a beautiful color is a bright color; Plotinus simply explains here that its beauty *is* its brightness. Plotinus makes this statement in the course of explaining that the true cause of beauty in bodies is form, which "fits together from many parts something that is one by composition," and so causes the beauty of proportion but also, as light, causes the beauty of color. These two visible manifestations of form have very different relations to the body. Proportion is the beauty of bodies as bodies, or of the embodied activities of the soul. Bodiless forms, since they are simple, have no proportion and do not share in this kind of beauty. The beauty of light, on the other hand, is the beauty of the bodiless, since light itself is bodiless and simple. The more a body resembles light, the more it reveals the characteristics of the bodiless forms. This, Plotinus explains, is why "fire itself is beautiful beyond other bodies, since it holds the rank of form in relation to the other elements. Its position is above them, and it is finer than other bodies, since it is close to the bodiless."[64] Though Plotinus describes colors and fiery bodies as dematerialized, he almost never describes visible places this way, though he does say once: "every garden is a radiance (ἀγλάισμα)," meaning to play later on the similarity of the word "radiance" (ἀγλάισμα) to "statue" (ἄγαλμα).[65] We can see here the potential for Vincent Scully's characterization of Neoplatonism as a method of dematerializing bodies. Were Plotinus to take an interest in seeing the intelligible manifested visibly in a place, he would undoubtedly promote places that manifest dematerializing characteristics like light, which attach to the natural forms of things without the need for an added symbolic structure. But Plotinus never suggests that we should seek out such places, and he does not offer a method for creating them. He never suggests that we should dematerialize anything but our own soul, which is already properly immaterial.

We find one other intelligible characteristic that may be manifested directly in the visible realm, and which clarifies the nature of dematerialization. Plotinus uses the term "stillness" (ἡσυχία) to describe the activity of beings that efface their own characteristics in order to acquire intelligible characteristics. When a body becomes still, it ceases to reveal its character as body, and can reveal the silent contemplation of the intellect. Souls, too, become still, when they want to approach more closely the character of intellect. Plotinus says once: "the soul held itself in stillness (ἡσυχίαν), yielding to the activity of the intellect."[66] He attributes the characteristic of stillness even to the universal intellect and the One, but the term means something different for them than it does for souls and bodies. While the stillness of souls and bodies is a departure from their proper activity, "the stillness of the intellect is not an effacement (ἔκστασις) of the intellect. The stillness of the intellect is an activity which maintains a respite from other things."[67] The intellect's stillness is its remaining in itself. The One has stillness in yet another sense. It must refrain from becoming anything—it must remain still—if it is to be responsible for all things.[68] This last use shows that stillness differs slightly from beauty. Plotinus usually wishes to deny that beauty has anything to do with the One, restricting its highest form to the universal intellect. Stillness, on the other hand, does belong to the One, so it is not an intelligible characteristic in the same way that beauty is.

Plotinus devotes one of his rare evocations of sacred places to this characteristic. He advises the particular human soul to look at the universal soul, the "great soul," which is "established in stillness."[69] Plotinus then advises us to observe certain characteristics of the body of this great soul: that is, the visible cosmos. He says: "let not only its surrounding body and its tides be still, but also everything that surrounds it. Let the earth be still, the sea be still, the air and the sky itself untroubled." Plotinus invokes the stillness that comes over the four regions of the cosmos: the earth, water, air, and fiery heaven. He invites them to be still so that they may more closely resemble the soul that he is about to pour into them. We pay attention to their identity with soul rather than their difference from it, so that we can then see how soul creates the world in stillness, as Plotinus himself says elsewhere.[70] He invites us to consider "soul as everywhere pouring and flowing from outside into it, while it is at rest."[71] He does not here describe an experience he has had of the world outside him, but an exercise within the mind, which may help us understand the relation between the cosmos and the universal soul which informs it. Though the cosmos we visualize is imaginary, the stillness he asks us to envision is not. We let the world as visualized inside us acquire real stillness so that we can more clearly discern the work of the universal soul in us as well as in it.

Stillness differs from the beauty of proportion, but resembles the beauty of light. The beauty of proportion characterizes bodies *as* bodies. A body can be beautiful simply because of the proportion of its parts, without directly revealing the intelligible. But to the degree that the body resembles light made visible, it takes on the incorporeal and intelligible characteristics of light and dematerializes itself. Stillness, too, allows a body to take on the characteristics of the intelligible. When a body becomes still, it dematerializes itself, suppressing its character as an image so as to reveal directly the positive presence of its intelligible source.

The Neoplatonic Landscape

Plotinus may call only for an interior visualization of stillness in the world, but the Neoplatonists also follow the tradition of seeking out stillness in exterior landscapes. Pierre Hadot has pointed out that the writers of antiquity favor two quite different kinds of landscape: the "charming landscape," and the "grandiose or sublime landscape."[72] Hadot identifies the ancient archetype for the charming landscape as the grotto of Calypso from the *Odyssey*, but he notes that in general the components of the landscape "are always woods of various kinds, birds, meadows, a murmuring spring, flowers, fragrances, a breeze, fruits, the pleasure of all the senses." He also stresses the solitude that accompanies the charm of the landscape, and uncovers several examples of it in the writings of the Neoplatonists. We find the best of these examples in Iamblichus' explanation of the morning activities of the Pythagoreans. Chief among these activities was a walk. He tells us that the Pythagoreans "took their morning walks by themselves and in such places where calm and stillness were appropriate, where there were temples and sacred groves and any other pleasure for the spirit."[73] The Pythagoreans sought out a place that was both still and sacred, and

Iamblichus explains why: "they thought they should not meet with anyone before they established their own soul and set their reasoning in harmony." Iamblichus here ties together the three qualities of stillness, sacredness, and thought. The exterior landscape of the temple or the sacred grove has a stillness about it, and because of this stillness, it is suitable for the cultivation of the soul. That is to say, the intellectual characteristic of stillness manifests itself sensibly in the sacred precinct, and this fosters the interior stillness and establishment of the soul within the Pythagoreans. Because of this unusual characteristic, the Pythagoreans did not merely take their morning walk in the sacred precincts, but tried to be always in some such place: "after their morning walk, they then met with others, but mostly in the temples or, if not, in similar places."

Hadot gives no examples of "the grandiose or sublime landscape" in the writings of the Neoplatonists, but we find one in Damascius' *Philosophical History*. Damascius, who traveled the Near East with fellow philosophers, records in this work his impressions of the river Styx as it flows past the town of Dion in Arabia. The river descends into a canyon here, whose bottom must have been significantly wider than the river itself, for Damascius records the presence of "gardens and many farms" here. The focus of Damascius' description lies at the canyon's far end, where it narrows, and the Styx cascades over a great waterfall, producing clouds of mist that dissipate into the air and collect again in the pool at the bottom. Damascius comments: "this vision (θέαμα) and work of nature is august (σεμνόν) and terrible. There is no man who, seeing it, would not be filled with reverent (σεβασμίος) fear."[74] The term "vision" does not necessarily denote a religious sight, but the following two terms—"august" and "reverent"—clarify that the waterfall has a religious significance. Struck first of all with the religious effect of the waterfall itself, Damascius goes on to describe briefly the local cult that has arisen around the "work of nature." The locals offer gifts to the divinity, which sink into the pool regardless of their weight when the divinity accepts them. The locals also swear on the region and its waters, with a grave divinely administered punishment for those who break such an oath: they die of dropsy within the year.

Damascius' description of the Stygian waterfall has the character of a late antique tour guide about it. His editor, Polymnia Athanassiadi, has aptly described his travels in the region as "spiritual tourism."[75] He does not provide instruction for participation in the cult, a participation which the locals already know and which the pilgrim should learn from them. He rather narrates the physical characteristics of the place, as well as the local color, and his affective experience of the waterfall. Pierre Hadot has described the character of this kind of experience: "the desire for travel and diversion often mingles with the religious need in the psychology of the antique pilgrim. One moves rapidly from pilgrimage to tourism, from fervor to curiosity."[76] To this desire for the satisfaction of an exterior place, Hadot opposes the philosopher's preferred emphasis on an interior journey. The symbol of this interior state of mind is the very bucolic landscape with which we began, a sacred place that is "less an objective place than a state of the soul projected onto nature."[77] This state of the soul may be embodied in an objective place, like the countryside of the pastoral poets, or it may be visualized as a landscape within the soul, as when Plotinus visualizes the quiet cosmos within the soul, so that the soul may experience the tranquility it offers. He famously advises us to "flee to the beloved fatherland," but the fatherland we seek is within us, not a place outside us. The exterior place is, at best, an image of an interior reality.

Plotinus: The Soul's Salvation Needs No Place

Many of Plotinus' students held the rites of the Hellenic religion in much higher esteem than Plotinus himself did. His student Porphyry wrote a treatise on Hellenic religion, and included in his biography of Plotinus many details of religious life in Rome that are not indicative of Plotinus' own concerns. Porphyry tells us that his fellow student Amelius took to making animal sacrifices and celebrating feast days at the local temples. On one occasion, Amelius asked Plotinus to join him, but Plotinus replied with what has now become a famous rejoinder. The gods, he said, "ought to come to me, not I to them."[78] Porphyry notes that he was mystified by this response, as were his fellow students. It has produced a similar quandry among present-day scholars.[79] At the very least, we may say that Plotinus' attitude here reflects a real tension between Hellenic religion and Neoplatonic philosophy, and we find it played out in various venues over the course of Late Antiquity.

The young Julian encountered such a tension when he came to study philosophy with Aedesius at Pergamon. Aedesius felt himself too old for the task, and recommended that the boy study with his student Eusebius. We are told that Eusebius was a master at "sharpness in the parts of an argument, dialectical snares and nets."[80] Eusebius was frequently joined by his fellow student Maximus, who preferred the study of miracles and divination. To warn the young Julian off of Maximus, Eusebius would close each of his exegetical lectures with the statement: "these are the real beings, while the witchcraft and magic that deceive sensation are the work of wonder-workers, who go mad and yield to certain material powers." Eusebius related an occasion when he joined Maximus at the temple of Hecate for a demonstration of Maximus' power. Eusebius, along with the others who had joined him, sat down, while Maximus stood in front of the statue of Hecate. As Maximus burned incense and sang a hymn, the visage of Hecate seemed to smile, and then to laugh. Shortly, at the word of Maximus, the torches in the hands of the goddess burst into flame. Eusebius conceded that he and his companions left the place astonished by Maximus, "that theatrical worker of wonders." But Eusebius concluded his story with a caution: "you must not wonder at such things, just as I do not wonder, but understand that the great thing is the purification that comes about through reason."[81] Eusebius opposes his own rites of the mind to the rites of the body, at least as Maximus displays them. Eusebius seems to have lost his war for the soul of Julian, as Julian is recorded to have become a follower of Maximus, and to have attempted to reinstitute a living form of Hellenic rite within the Roman empire.

Plotinus does not neglect the religious practice of Hellenic religion in order to confine human endeavor to the discursive work of reason and the analysis of the visible world. He is no secular humanist *avant la lettre*. He simply finds religious practice unnecessary for ascent to the divine, having found in his own soul a more direct route. He recalls his experience of this route in the only explicitly autobiographical passage in the whole of his work: "often, awakening into myself from the body, I have come to be outside of other things, but inside myself, seeing a wonderfully great beauty. I believe then that I belong most of all to the greater part, and exercise the best life, and come to identity with the divine."[82] Plotinus claims here that he acquires the divine life not through any action or disposition of the body, but through a "waking up" from the body, which allows the soul to enter the realm of "supreme actuality"—the divine mind. Such an abandonment of the body once earned Plotinus a reputation as a gnostic. That reputation has been shown to be undeserved, but not because Plotinus thinks we belong essentially to the body.

Plotinus wraps his psychology around the insight that there is a "better part" to which I as a subject may belong. This "part" is his famous undescended soul. He claims: "not all of our soul descended, but there is always something of it in the intelligible."[83] We are not always aware of this part because the other part—the one "in the sensible"—"does not allow us to have sensation of what the soul above contemplates."[84] We are not identified with either part of our soul, but the lower part of our soul can take us captive, as it were, and prevent us from seeing what our higher soul is doing. When this happens, we only sense what our five exterior senses perceive, and are aware only of bodies and not the intellect. Our higher soul is still operative even when we are not directly aware of it. In fact, we could not think at all unless the intelligible forms perceived by the higher soul were handed down to the lower soul so it could apply them to the data received through the senses. As Plotinus puts it, "what has been intellected comes into us when it descends as far as sensation." We sense visible objects directly; we sense intelligible objects indirectly, by their descent from one part of the soul to another. If we could cease to identify ourselves with the lower part of the soul, and awaken into ourselves, as Plotinus claims to have done, then we could see the intelligible directly. Seeing the intelligible means being the intelligible—since at the level of intellect the subject contains its object—and so our ability to see the intelligible means that a part of us must in some way be an intellect.

As we have already seen, Plotinus advocates an interior transformation of the soul to produce awareness of the soul's own higher part, and so we do not need to explore or manipulate the visible world to restore our contact with it. The temple and its visible rites do not save the Plotinian soul. The soul itself is the temple of the Plotinian system. Following a well-established Platonic tradition,[85] Plotinus uses the physical structure of the temple as a metaphor for the interior rites of the soul in his treatise "On the Good or the One."[86] The temple metaphor appears in the last pages of the treatise, and so, in Porphyry's edition, which puts this treatise last, it constitutes the conclusion of Plotinus' life work, even though Plotinus wrote it relatively early in his career. Plotinus has been speaking in this treatise of the union between what is seen and the one who sees, in the special case where someone attempts to see the One. Such a person is like "someone who goes into the inside of the shrine, leaving behind the visages in the nave." The nature of these visages is unclear, largely because Plotinus does not specify whether he is describing a Greek or an Egyptian temple.[87] A Hellenic temple may have statues of the gods in its nave, while an Egyptian temple would be more likely to represent the gods with engravings carved on its walls. Whether statues or engravings, these visages are "things to be contemplated" (θεάματα). Within the sanctuary, however, there is not contemplation, but "another way of seeing: self-effacement (ἔκστασις), simplification, yielding of oneself, reaching for contact, rest, and an indirect thought that makes one suitable for it." We can use the language of sight only metaphorically to describe what is in the inner shrine, since there is not properly anything to see there. In this metaphor of the temple, the soul's higher part plays the role of the nave. There, the soul has in itself the visages of the first principle: the intelligible forms of all things. Beyond the nave, the soul can enter something beyond even its higher part, something which cannot properly be described in language at all. None of these activities needs the rites of the Hellenic religion and the visible temple in which they take place. The places that are sacred to the soul are within it, in its own parts, and afford the place of interaction even with the One, a good that is higher than soul and intellect themselves.

Critique of Plotinus: the Soul Requires a Sacred Place

Plotinus seems to have realized that his doctrine of the undescended soul would be controversial. He introduces it somewhat defensively: "and if I must dare to say more clearly how it appears to me, against the opinion of others, not even all of our soul descends."[88] Plotinus does not say who his opponents are here, but many later Neoplatonists are critical of this aspect of Plotinus' thought. Iamblichus claims that Plotinus' account of the undescended soul matches neither our experience of ourselves nor the way we treat other human beings, criticisms to be echoed and expanded by Proclus and Damascius. Carlos Steel has argued that this overt conflict between Plotinus and the later Neoplatonists is more rhetorical than substantive,[89] but it does shape the way these later Neoplatonists view the role of place in the soul's salvation, and so I will take it seriously here. In his commentary on Plato's *Timaeus*, Iamblichus makes several arguments to prove that the undescended soul doctrine makes our experience and treatment of each other incomprehensible.[90] First, the undescended soul doctrine requires that part of us remain outside the bodily, while our mixed soul—mixed of a rational part and an irrational part—is stained with the bodily. If this is the case, then who errs when the soul breaks the law in order to follow its own passions? The irrational part cannot err, since it has no ability to choose anything other than following its irrational desire. Our actual treatment of someone who has erred provides a clue. When we say that someone has erred, we blame their choice, not their desire. Choice, however, is rational, and so implicates the mind in the error of the whole person. If the whole person has made the mistake, then clearly no part of it has remained in the realm of intellect. Iamblichus' second argument begins with the observation that, according to the doctrine of the undescended soul, the best part of us—our intellectual part—is in constant possession of the truth, and so also is happy. If this is the case, "what prevents all of us human beings, even now, from being happy, if our summit always intellects and is always directed at the gods?" In fact, we are not always happy, and so we cannot have a part of ourselves always in contact with intellect. Iamblichus' argument here requires that the soul behave as the unity of its parts. If one part remains undescended, then the soul as a whole must be aware of this. There can be no absence of awareness such as Plotinus attributes to the soul's lower part in his treatise on the soul's descent. Iamblichus' last argument relies on the authority of Plato, who in his *Phaedrus* describes the soul as a chariot, governed by a charioteer and driven by two horses. Plato describes the chariot as winged at the best of times, orbiting the earth with the heavenly bodies, while at the worst of times it loses its wings and falls to earth with its charioteer and horses. The charioteer, Iamblichus says, represents the "most graceful and, so to speak, the most principal part of us." If the charioteer is our highest part and it falls to earth with the rest of the soul, then Plato clearly does not think that we possess an undescended soul.

The fully descended character of all human souls does not mean that they all interact in the same way with the intellect that now transcends them. In his major surviving work, *On the Mysteries of the Egyptians, Chaldeans, and Assyrians*, Iamblichus takes pains to distinguish the rites of the sage from the rites of the masses. "The great herd of human beings," he says, "is ordered under nature."[91] His comparison of the masses (*οἱ πόλλοι*) of humanity to cattle does not intend to denigrate them, but to show how their state demands a certain form of rite. The state of the herd is the state of nature. It is "governed by natural powers," "looks down toward the works of nature," "fulfills the governance of fate," and "always uses practical reasoning concerning only the things of nature." The repetition of "nature" here cannot help but stand out as the defining characteristic of the great herd, as it is repeated in nearly every clause. Behind the term "nature" lies a certain way of life: governed by the motion of the heavenly bodies, from which all earthly motion derives, and dedicated to the reproduction of the species, which is the way an earthly body imitates the unceasing motion of the heavens. Human motion here distinguishes itself from other animal motion only because it uses reasoning, and not mere instinct, to achieve the reproduction that is its goal. The great herd of humanity, then, directs itself to a life of survival, growth, and reproduction, which differs little from that of other animals.

Contrast this with the "certain few" who stand outside the herd. These few "make use of a certain supernatural power of mind," "are removed from nature," and "come around to a separate and unmixed mind."[92] These last two terms, "separate" and "unmixed," traditionally characterize the divine mind. Were Iamblichus to claim that a "certain few" can achieve the timeless thought of intellect, he would misunderstand the nature of the soul just as he thinks Plotinus has. Iamblichus' "certain few" instead make use of a human intellect and unity, images of the divine intellect and unity.[93] Most of us put our intellect and unity into action only through the medium of soul. To utilize them directly is "the rarest of all things," and "comes about in one person hardly ever and late in life, in the completion of the sacral."[94] Iamblichus goes on to deny that we should propose the direct use of intellect as a goal to everyone, or to most of the participants in theurgical rites.[95] The care they exercise in the rites should be bodily in form. In fact, we need not even propose the direct use of intellect to someone capable of it, since "our argument does not lay down customs for such a man, for he is greater than any law."[96] In other words, anyone able to interact with the highest of the gods through an immaterial mind can have no one above him to give him counsel. He instead becomes the standard for others. Iamblichus may have Plotinus in mind here; Proclus will later ascribe such characteristics to the fictional Parmenides in Plato's dialogue of the same name.[97]

Those who do not possess a supernatural mind need not worry that they will be lumped with the great herd of humanity. In accordance with his guiding principle—that between every two extremes there is a mean—Iamblichus identifies for us a "middle class," which blends characteristics of both sage and herd. We do not hear much about the special characteristics of this class. Instead, Iamblichus explains the three ways in which it arises. The first way to join the middle class is to follow both nature and intellect. We may imagine Socrates as a member of this class, as he has both a wife and children (following nature), but also engages in the exercise of intellect to such a degree that it is easy to read Plato's dialogues on him without realizing that he has a family (following intellect). One may also "participate in a sort of life which is mixed from both." This way of life differs from the first in that the work of nature and the work of intellect are not now conceived as separate. Such a way of life collapses the distinction between body and mind in favor of human faculties that combine the two. Third, one may be in a process of abandoning the lower form of life for the higher. While the first two forms of the middle class are sustained by nature, this last form aims at its eventual abandonment, and is suited to those who aspire someday to be sages. Since sagehood requires a lifetime, the aspirant must remain in the middle class until his transformation comes to pass.

It is no small matter to know which class we belong in, because the different ways of life demand different rites. The great herd of humanity, which follows a life of nature, employs "a worship suitable to nature and the bodies moved by nature."[98] Such a worship will make use of bodies, and for Iamblichus, this necessitates the use of places as well. He says that such worship will employ "places, air, matter and the powers of matter, bodies, the dispositions of bodies and their qualities, fitting motions and the changes of things that come to be, and everything else possessed by such things." The sacrificial rite performed by a citizen of Athens, for example, is not performed just anywhere. The citizen goes to the temple of Apollo at Delphi, for example, and here the priest, after perhaps regarding the celestial and atmospheric omens, takes the body of an animal, performs certain motions over it, and, after slaying it, takes into account the disposition of its entrails and their qualities. All of the particular characteristics of the sacrifice—the ritual motions, the body of the animal, the reading of the entrails—come to pass within the bounds of the place, which is itself the largest visible component of the rite.

As we might expect, the sage engages in a form of worship that is precisely the opposite of the herd's. Iamblichus says of it only that it is the "intellectual and bodiless custom of the sacral." He does not say anywhere in *On the Mysteries* what this custom is, but he hints at its nature when he later says that the bodiless rite will cultivate the gods with offerings of "virtue and wisdom." The sage offers his faculties in place of the slain animals offered by the herd. Iamblichus' text tells us no more than this, but Gregory Shaw has speculated that the bodiless rite involves a "theurgy of numbers."[99] He relies primarily on Neoplatonic references to the Pythagoreans for hints as to what a Neoplatonic theurgy of numbers might look like. Iamblichus tells us that Pythagoras taught a "divination through numbers,"[100] while Proclus says that the Pythagoreans "consecrated numbers and geometric shapes to the gods."[101] These uses of number do not replace theurgic rites with mathematics in order to accomplish the salvation of the soul. Instead, mathematics becomes a form of theurgy, though one suited only to those who have completely freed themselves from attachment to nature and the body.

What rite does this leave to the middle class? Again, Iamblichus gives no special characteristic to the median between the bodily and bodiless rites, but he suggests different ways of sharing in both.[102] The married Athenian scholar outlined above may make sacrifices at the temple of Apollo, but in the courtyard of the Platonic school he offers the gods his own virtue and wisdom. Or one may use both together as "the foundation of worthier things." This form of worship mixes the bodiless and the bodily rite, perhaps by combining the performance of actions with an intellectual contemplation of them, as Dionysius the Areopagite will later prescribe for the Christian laity. The better things, in this case, are the heights beyond both the body and the intellect: the unity of the gods themselves. Finally, one who hopes one day to merit the bodiless rite alone may remove himself from the bodily rite, and set himself on the path to the bodiless rite. He adopts an ascetic practice, devoting himself to the bodily rite out of necessity, but putting all his effort into the development of his mind.

The threefold structure of human ways of life and worship so far seems to be a result of the varying constitution of human beings. Not everyone is made the same way, and so not everyone can worship the gods in the same way. But Iamblichus refuses to make the three structures of worship solely dependent on the varied characters of human beings. He posits a threefold character to the gods, so that each mode of worship, in addition to its suitability for a certain kind of human, also accords with a certain kind of god. Some of the gods act on the world we see through the intermediaries of the heavenly bodies: the sun, moon, stars, and planets. Iamblichus says that these gods "have a soul and nature underlying and serving them for their creations, to the degree that they wish."[103] When we look up into the night sky, we see these gods, acting through their visible nature, the luminous bodies of the heavens, to move the cosmos according to their will. The motion of the heavenly bodies finds its way into the motion of earthly bodies, and ultimately into all motion on earth. Not all gods undertake this continuous creation and recreation of the cosmos. Some remain outside it, in the sense that they are outside extension in mass. Iamblichus says that these "are entirely separate from soul and nature."[104] They do not appear to us at all, and must remain mysterious to anyone who does not have direct intellectual access to the divine. Midway between these is a middle class of divinities who "supply a communion of the one with the other." These divinities do not act through the heavenly bodies, yet they are not separate from them either, and so they link the separate gods with the embodied gods. Iamblichus does not seem to be drawing on a preexisting tradition here, but on the mathematical necessity of such gods (given that there must be a mean between any two extremes), and he says no more about them.

The middle class of gods receives the most complicated of the gifts prepared for sacrifices. While the separate gods receive the intellectual gifts of virtue and wisdom, and the heavenly gods receive natural bodies (such as animals), the median gods receive either: (1) twofold gifts; (2) gifts which have something in common with both forms of divinity; or (3) gifts that rise from the lower to the higher. In other words, the middle gods may be cultivated in the three ways common to the three middle ways of life, and the three forms of worship. In all cases, the object is to offer to the gods the proper subject of their government, so as to make explicit the bond between governor and governed.[105]

Critique of Plotinus: the Gods Require a Sacred Place

Iamblichus takes up the task of articulating the way in which the gods require material rites in *On the Mysteries*, particularly with reference to the action of sacrifice. Porphyry, in the letter to Iamblichus that motivates the writing of *On the Mysteries*, has raised two objections to the rite of sacrifice. The sacrifice of meat to the gods seems incompatible with the vegetarianism required of the priests who perform the sacrifice, an objection that does not concern our present task. But Porphyry's second objection—that the sacrifice has no clear utility—prompts Iamblichus to explain why the gods manifest themselves in the specific places of the sacrifice. Iamblichus opens his discussion by ruling out a number of possible reasons for making sacrifices. We should not make sacrifices to honor the gods, to thank the gods for the gifts they have given us, or to pay back our debt to the gods for such gifts. Such reasons for sacrifice treat the gods as though they were simply powerful human beings who deserve honor and who shower us with gifts. We perform ritual acts of thanksgiving and debt repayment to such human beings because we could not possibly repay them in kind. The emperor of Rome, for example, merits sacrifice not because he is a god, but because he takes care of the average citizen in a way that the citizen cannot possibly repay. The gods, however, are not emperors. To treat them as though they were emperors "does not at all preserve the transcendence of the gods and their rank as removed causes."[106]

The sacrifice may also be misunderstood as a work of nature. As the Neoplatonists understand it, the cosmos holds together through a thread of connections that bind each thing to every other, a "sympathy" (συμπάθεια) which arises because the cosmos has a single soul, the elder sister of our individual souls. Through this cosmic soul, we are in touch with both the cosmos and each other in a way that transcends bodily contact. This doctrine is not mystical and esoteric. It belongs not to the Hellenic priesthood, but to natural science. Plotinus, for instance, relies on it to explain how sense perception occurs. I am able to see things around me because the thread of unity binds me to them in such a way that I am both the same as and different from them. If there were a body outside the cosmos, I could never see it, because my sense perception of other things depends on our common filiation with the cosmic soul.[107] The thread of cosmic sympathy runs deeper than the manifest sympathy we experience in sense perception to include connections that require initiation in the mythic context of the Hellenic rites before they can be discerned.[108] Heliotrope has a special connection to the sun, the ram has a special connection to Athena, and so on. These connections—having the nature of symbols rather than images—determine the things to be used in sacrifices. Because the cosmic sympathy assigns a specific animal to a particular sacrifice, some worshippers assume that sympathy causes the effect of the sacrifice in the way that sympathy between myself and a tree causes my sense perception of the tree. If this were the case, the sacrifice would be a work of nature just as is sense perception, albeit accomplished by less obvious channels. Iamblichus tries not to offend such an initiate too deeply, saying that this view "says something of what is true, and what necessarily follows upon the sacrifice," but it does not reveal "the true way of sacrifices."[109] To ascribe the efficacy of the sacrifice to cosmic sympathy mislocates the nature of the gods. Just as the first misconception thought of the gods as like emperors but more powerful, the second misconception treats the gods as identical with nature. The substance of the gods, however, "does not lie in nature and natural necessities so that it could be roused by natural passions or powers extended though the whole of nature." While natural sympathy determines the choice of things used in the sacrifice, it is only an "accompanying cause" (συναίτιον), and depends on "preceding causes."[110]

The true causes of a sacrifice's efficacy are "friendship," "familiarity," and "a relation that binds the created to the creators, and the begotten to the begetters."[111] Iamblichus explains how these terms work in the case of the material gods, the ones that have the planets as their bodies, and who "have a kind of communion with matter inasmuch as they preside over it."[112] Their close relationship with matter allows them to preside over material processes like "division, striking, rebounding, change, generation, and corruption of all material bodies." If we wish to cultivate such gods, "we must offer them a cultivation which is material, since they are material. We are thus brought wholly into familiarity with the whole of them, and we bring to them in our cultivation what is fitting and of their kind." Following the principle that we unite ourselves to other beings by offering them what is "of their kind," Iamblichus concludes that material sacrifices bring us into familiarity with material gods.

We should not assume that the material gods are in the heavens, and that we sacrifice to them from a distance, hoping that the smoke from the sacrifices rises high enough to reach them. Porphyry wishes to keep the gods in the heavens, radically removed from place in its earthly aspect—the kind of place we as human beings can move through. The priests, on the other hand, often call upon gods as though they inhabited places on earth or beneath the earth, and so Porphyry pointedly asks for clarification in his letter. Now Iamblichus himself does not want to say that gods are literally on or beneath the earth, since this would limit their power to certain places within the cosmos. He has already shown that, if the gods are to have the characteristics we attribute to them, their power must extend always to the whole of the cosmos, never to a part. And yet "all things are full of gods," as the ancient proverb goes, and the priests are right to call upon them in places above, on, and beneath the earth. Iamblichus attempts to provide a philosophical precision, which allows the gods to be invoked as in a place without their power being limited in any way: the gods enter places not in their substance, but through an "allotment" (λῆξις).[113] Although Apollo, for example, exercises power over the whole cosmos, he receives as his allotment certain parts of that cosmos. He may be allotted one of the four great regions of the cosmos—the earth, the sea, the air, or the fiery heaven—or sacred cities, earthly regions, certain temple precincts or sacred statues.[114] Apollo exercises power in place, from the largest order—the cosmic regions—to the smallest order: the surface of the body or, in this case, the sacred statue. He does not, however, enter into any one of these places in the way that human beings do. Just as the sun illuminates all things with its rays, while itself remaining outside of them, so the gods illuminate all things from outside of them. The light of the sun enters into all things, but it does not mingle with them, as we can see when we draw the shade on a window, and the light that had pervaded the room does not remain on the things it once illuminated (as would be the case if it mixed with them). It vanishes. The gods, too, enter into all things in order to illuminate them, but they do not mix with any of the things they illuminate. Their illumination cannot, then, be broken into pieces, with one piece for each sacred place. The presence of Apollo at Delphi is the same presence at Colophon.

The identity of the god's presence at each sanctuary does not mean that the particularity of the sanctuary can be abandoned in favor of a ritual that simply worships the entire cosmos at once. The gods which govern the cosmos are different gods, and though each presides over the whole, they manifest themselves in diverse ways owing to their own multiplicity. The gap between their presiding over the whole and the particular places in which each manifests itself is filled by the intermediary orders of the daimons and the heroes. These "participants of the gods are such that some participate etherially, some aerially, some aquatically."[115] Some daimons bring the presence of the gods to fire, air, and water, but some bring it also to smaller regions and places. In every case, the heroes and daimons administer the part with the light of the god which that part is adapted to receive. Apollo, then, shines a single light over the entire cosmos, but only certain places within the cosmos are adapted to receive that light. The mythic context of the Hellenic religion privileges these places, and they become the cities and countries under his patronage, and his temple precincts.

What is it about certain places that makes them adapted to receive the illumination of the gods? Iamblichus explains: "since it was necessary that things on earth not be entirely without a share of divine communion, even the earth received a certain divine share from it, enough to make space (χωρῆσαι) for the gods."[116] The "divine share" here is not the actual illumination of the gods, but the natural capacity to receive that illumination. The priests do not create this capacity; they discover it through the mythic context of the religion, and then make use of it in their rites. The mythic context reveals the hidden sympathies between earthly things and heavenly things, and ultimately between earthly things and the gods, and so also reveals "containers fitting for each of the gods, by their familiarity." These containers may be types of stone, animals, plants, and incense, so long as the matter has the three characteristics of being "holy, perfect, and deiform." The theurgists then "work a complete and pure container from all of these." Iamblichus shortly explains what he means by this distinction between the thing as container in itself, which requires only discovery, and the thing as component of a larger container, which is "worked" by the theurgists and is more complete. He says: "we must choose matter that is familiar with the gods, since it is able to attune itself to the dwellings of the gods, the installation of statues, and the sacred works of sacrifices." The larger container is not the stone or plant, but the sacred places and their rites, both of which make use of particular sacred things in their structure. Since the sacred things are familiar to the gods, they are able to become part of the symphony of things that constitutes the sacred place and its rites. Without this foundation of sacred things, no "participation in the reception of higher beings can come about by places on this earth or the people who dwell here." The things used in the rite, while they have in themselves a hidden sympathy with the heavens, and so are always sacred in a certain sense, only fulfill their sacred character by becoming part of a "complete container." Only then is their "image" character thoroughly replaced with the "symbol" character. Athena does not enter into any ram, but the ram which has been prepared as part of a "perfect container" for the gods, a container comprising the temple precinct, the temple building, the tools used in the rite, and the motions of the rite itself.

Distinction and Opposition of Image and Symbol

"All things are full of gods"—the continuing influence of this ancient proverb, and its philosophical interpretation by the Neoplatonists, prevents a sharp split from arising between the sacred and the profane, between the temple and the wheat field. The ram may only become a symbol as distinct from an image in the course of the rite, but its symbolic character remains to some extent even on the hillside, where it primarily manifests its natural characteristics, that is, its image of the divine form.

In one case we do find Iamblichus making a sharp distinction between sacred and profane. When the gods govern a particular animal, plant, or anything else on the earth, they give that thing a special significance. It "participates in their authority, and grants to us an indivisible communion with them."[117] Some things familiarize us with the gods when they are sacrificed by being killed and consumed by fire. Others, however, produce this familiarity only by the opposite course of action: the preservation of their lives. Such things, "when they are preserved and guarded, increase the familiarity between those who keep them and the gods. By remaining unharmed, they preserve the power of communion between gods and human beings." Iamblichus refers here specifically to the Egyptian practice of forbidding certain animals to be sacrificed or harmed in any way, a practice described at length by Herodotus.[118] In Herodotus' account, those who kill such animals deliberately are executed, while those who kill them accidentally must pay a fine. Such regard for nonhuman living things arises not from their "intrinsic value," but from their capacity to receive the illumination of a certain god. The pious Egyptian sees the gods as present in the animal, since the animal has a place in a larger context than one directed simply at human use, and he preserves the animal within that larger context.

A certain kind of advantage at least partially motivates the traditional respect for sacred animals, plants, and inanimate things: the fear of divine wrath. Iamblichus eliminates this element from his system, explaining that what we call divine wrath is actually an action on our part, "a turning away from the care of the gods which works our good. When we turn ourselves away, it is as though we hid ourselves from the light of noon, bringing darkness on ourselves and depriving ourselves of the good gift of the gods."[119] The common conception of angry gods is thus transformed into a conception which preserves their freedom from passion, while accounting for the traditional stories of divine wrath. Hand in hand with this rethinking of wrath goes a new motivation for respecting sacred things—not fear, but desire for the conscious reception of the "good gift of the gods." Iamblichus attempts to free the rites from both an incorporation of the gods into a purely human system of action based on reactive emotion (fear), and the parallel tendency to reduce the rites to an operation of natural science. When the gods are treated as comprehensible within the realm of human experience, owing to their all-too-human behavior, the rites cease to require their symbolic character. Symbols are necessary only for an interaction with the unknowable, while a powerful but knowable god could be placated or persuaded through the same means as human beings. These means could be identified on the basis of the natural forms or images of the things concerned.

While the Iamblichan rite necessarily distinguishes symbol from image, certain aspects of the theurgical art go further, establishing not merely a distinction but an opposition between the symbol and the image. The human arts of prophecy take this additional step, looking for breaks in the otherwise smooth surface of cause and effect in the visible world. Iamblichus deals with three such arts: they look at, respectively, animal entrails, bird behavior, and the motion of the heavens. He cautions us that these arts are intimately bound up with "probability and seeming," but does not directly say why these arts are less certain than the other arts of divination.[120] The reason seems to be that they do not involve a transformation of the human knower into a divine knower. Where oracular prophecy requires an ecstasy on the part of the prophet, these three arts are "human crafts," in that they allow us to seek out the divine without becoming divine. This is why these arts require an opposition between image and sign in their object. The ritual forms of prophecy, which are not human crafts, do not require that the image contradict the symbol, since the prophet must ascend beyond the visible world where both image and symbol have their place. Here, however, we need to see the presence of the intelligible revealed at the level of the visible, and so the symbolic appearance of the intelligible must stand out *against* its appearance as an image that reveals no special providence of the gods.

The gods produce these signs either "through nature, which serves them in generation" or "through generative daimons." "Nature" here refers to the soul of the cosmos, which usually serves as the means by which the gods generate all visible things. In its signifying capacity, however, nature can oppose its own generative capacity. Nature acts directly on the motion of the stars, so it is not difficult to see how they can produce signs. It is not quite so easy to see how nature can produce signs in the other two prophetic crafts: the interpretation of animal entrails and bird behavior. Iamblichus explains that there are three causes of signification in these two crafts: (1) the animal's own soul; (2) the motion of the air in its revolution around the earth; and (3) the rotation of the heavenly bodies. These act in addition to the daimon set over the particular animal in question. The motion of the eagle as it flies from its perch, for example, arises from itself, the air that surrounds it, and the motion of the heavenly bodies, perhaps the sun, in particular, since the sun is most closely associated with Zeus, and the eagle is the symbolic animal of Zeus. Likewise, one of the daimons who serves Zeus may also guide the motion of the eagle. From its motion, the human student of the prophetic craft may "conjecture his prophecy, reasoning his way to it from certain probabilities."[121] The aspiring prophet clears a field for his conjectures by distinguishing what things do by nature, and observing instances where the natural does not come to pass. Only after he clears this field can he apply his understanding of particular significations to conjecture the exact meaning of a particular sign.[122] How do we know that signs exist at all, and that not every movement of the bird comes out of the nature that created it as an image of its intelligible form? Iamblichus explains that nature, in its generative function, never acts to prevent life. If it did, it would at the same time be acting to generate and to destroy life. If we see apparently natural acts that destroy or impede life, we may look to those acts specifically as signs. Iamblichus says of entrails: "an indication that they are so changed is that they are frequently found without a heart, or deprived of the most principal parts, without which it is not at all possible for animals to be supplied with life."[123] Of birds he says: "the greatest indication is that birds frequently precipitate themselves to the earth and destroy themselves, which it is not natural (κατὰ φύσιν) for anything to do; but this is something supernatural (ὑπερφυές), so that it is some other thing which produces these effects through birds."[124] When a bird dives into the ground in what is, practically speaking, a suicide, it acts not according to nature (κατὰ φύσιν), but supernaturally (ὑπερφυές).

In this case, we have a direct opposition between the bird as image, acting according to its nature, and the bird as sign, acting outside of its nature.

The two relations of image and symbol—distinction and opposition—are not wholly different from one another. When a being acts in a way that is *meaningless* in terms of its nature, we see it acting as a symbol in distinction from its activity as an image; when a being acts in a way that is *impossible* for it in terms of its nature, we see it acting as a symbol in opposition to its activity as an image. But both activities are "above nature" (ὑπὲρ φύσιν), and so both stand together as differentiated from activities "according to nature" (κατὰ φύσιν) and activities "against nature" (παρὰ φύσιν) such as "the mania that accompanies diseases."[125] While impossible activities may be more striking than meaningless activities, neither of them holds a special privilege in the rites. The divinatory rites, in fact, make use of both.

The Oracles

Porphyry mentions three oracles of Apollo—at Claros, Delphi, and Didyma (called Brandchidae by Porphyry after the family that runs it)—which serve as places of divination. In his response to Porphyry, Iamblichus provides a priceless account of how the sacred place employs the different relations of image and symbol. I will pay special attention here to Iamblichus' most thorough example: the oracle at Claros, near Colophon. Vincent Scully gives a depiction of the structure of the sacred place: "the temple itself was a large hexastyle Doric structure with three gigantic figures: of Apollo, Artemis, and Leto (Latona) standing on a common pedestal in the cella. Under them was a vaulted cave, reached by two stairways from the pronaos, and designed as a labyrinth of seven turns. This groped its tortuous way to a sacred spring and the final cavern."[126] The sacred place is not located arbitrarily. Scully notes that the site orients itself toward the horned peaks of Mykale, and rests in an enclosed valley.[127] The site has mythic as well as landscape associations. The sacred spring, from which the sacred water flows, is said to have originated in the tears of Tiresias' daughter Manto.[128]

Iamblichus describes the rites that precede, accompany, and follow the drinking of the water by the priest[129] of Apollo. While we have no way of knowing whether Iamblichus ever visited the temple at Claros, or whether he describes the rites accurately, his description of them tells us a great deal about how he sees the role of sacred places in the soul's salvation. Before drinking the water, the priest fasts for twenty-four hours. He "withdraws by himself to certain sacred places, inaccessible to the masses, as he begins to enthusiaze."[130] The masses may enter certain parts of the temple, but the priest must avoid these parts so as to bring about his thorough isolation. Iamblichus explains the significance of this withdrawal: "through his departure and release from human affairs he renders himself unstained for the reception of God." Human affairs, directed at maintaining the body and sustaining the species, often through the preservation of a political community, are not suited to the reception of divinity. They manifest the body *as* body, and do not suppress its own characteristics so that it may become receptive of something higher. By means of fasting and isolation, the priest "possesses the inspiration of the god, shining into the pure seat of his soul, and there is for it a. . .possession and a perfect. . .presence." Removal from human affairs takes the place of the soul's removal from the body in Plotinus, and allows the Iamblichan priest to make his soul a "pure seat" for Apollo. The priest, then, even before drinking the water, has already achieved perfect possession and presence, by means of the inaccessible character of the sacred place and the purity of his activity in fasting.

The drinking of the water takes place only on "certain ordained nights," after many sacred works have already been performed. The priest does not reveal himself to those present, but drinks the water from the spring in seclusion and then announces the oracles. If we reflect on this lengthy series of nocturnal rites performed in a subterranean cavern, totally blocked from the open air by the seven turns of the labyrinth which leads to it, we begin to understand why Scully suggests that the site reflects the "rather spooky drama" he considers characteristic of Late Antiquity.[131] Because the priest is able to tell the future after drinking the water, Iamblichus is obliged to consider the water prophetic, but he wishes carefully to dispel any impression that the rite at Claros is a form of magic.[132] He admits that "the prophetic spirit seems to extend through the water," but he declares that the truth is otherwise. If the prophetic spirit extended through the water, it would "roam extendedly and partially" through the water that participates in it. That is, it could be divided into parts, and it could be manipulated by virtue of having entered into a specific place. Instead, it "supplies and illuminates the spring from outside, and fills it with prophetic power from itself." The symbolic character of the water remains entirely distinct from its nature as water. Even as symbolic, it only prepares the priest to receive Apollo. It never becomes Apollo: "the inspiration that the water supplies is not the whole inspiration of the god, but it itself only produces a fitness and a purification of the luciform spirit in us." Through the water, in other words, "we become able to make space for the god." Once the priest has received the god, he performs actions that violate his nature as a human being. He is "no longer of himself, nor capable of attending to what he says, nor perceiving where he is." Only by standing outside his own nature can the priest speak the language of being at the level of the bodily, that is, utter an oracle.

From Iamblichus' description of the oracle at Claros we see that the oracle requires a certain place, a certain matter, and certain actions. It requires a place that is "inaccessible to the masses," in this case, by virtue of being underground, and hidden particularly by the labyrinth that separates it from the open air. The remoteness of the place from human affairs allows the divine a foothold in it. Likewise, the oracle requires a certain matter, in this case, water. We might assume that its prophetic character comes from its origin in the tears of Manto, but Iamblichus does not refer to this; he only points to the result. Prophetic sayings arise from the drinking of the water, and so we know the water must be sacred. The water, however, does not itself manifest the god. Instead, it acts as a preparatory agent, working a "fitness and a purification" of our own spirit, which is then able "to make space for the god." Finally, the oracle requires certain actions, in this case, fasting and sacred works, which put the matter into play in the place in such a way that the priest becomes a "friend" and "familiar" of the god.

The oracle at Claros reveals elements of all the relations between image and symbol we have mentioned. When the priest withdraws from human affairs into the deep recesses of the shrine, he seeks out a place where his proper activities as an embodied being have no place. It is not surprising, then, that Iamblichus' Pythagoreans rely on walks in the temple precincts to compose their minds, for these precincts suppress the noisy, distracting character of most human places. Iamblichus does not mention stillness as a characteristic of the sacred place here, but the isolation of the place resembles stillness in its effect. So far, we see nothing of the symbolic character of the place; the symbol emerges in the sacred character of the spring. Water by itself possesses no intelligible characteristic, unless in certain pools whose very transparency invites comparison with the intelligible, but Iamblichus here does not mention any natural capacity of the spring to produce oracular pronouncements. It is a special gift of the divine to this particular spring; hence the association with Manto and her tears. If we go beyond Iamblichus, and recall the landscape features pointed out by Scully, we see that the temple, while a human construction, takes its place among several symbolic landscape features that are not of human design or control. They have their place in the mythic context that transforms the being of the Hellenic people in the performance of the rites. The symbolic character of the sacred place culminates in actions that defy the nature of the priest who presides there. No longer operating at the level of a human being, and in fact doing things that transcend his very nature, he tells the future.

Porphyry has mentioned two other famous oracles, Delphi and Didyma, and so Iamblichus doggedly applies the same formula to them as he does to Claros. Because the formula is the same, we need not give them the same extensive treatment. Both Delphi and Didyma are oriented toward landscape features, and both involve the retreat of the diviner into a secluded portion of the place: the cavern of Delphi, and the inner shrine at Didyma, which is located at the highest point of the temple. When postulants enter the temple, they find that the entrance to the inner shrine, though immense, stands above a step that cannot be ascended by a human being. The postulant must instead descend a dark staircase into a walled-in laurel grove on the other side of the shrine. Both the diviner and the postulant, then, seclude themselves within the sacred precinct. The rites at Delphi and Didyma also involve a certain matter, which becomes capable of bearing something spiritual, though not the god himself. At Delphi, "an attenuated and fiery spirit" appears from the mouth of the cavern, and the prophetess "entirely gives herself up to a divine spirit, and is illuminated with a ray of divine fire." Iamblichus here seems to be influenced by the theory, popular in his time, that the prophetess at Delphi is put under the spell of volcanic fumes that emerge from fissures in the cavern. Walter Burkert notes that this has since been disproved geologically, and that "the ecstasy is self-induced."[133] At any rate, Iamblichus cautions that the god "is other than the fire, the spirit, the proper seat, and all the natural and sacred furniture that appears about the place." At Didyma, too, there is a "spirit which rises from the spring," but it is not the god. In the spirit, "another more ancient god, who is separate from the place, shines forth to the view, and who is the cause of the place, of the country, and of the whole divination." Finally, the rite requires certain actions, which allow the material to become symbolic within the confines of the place, though in neither case is Iamblichus sure that he knows what they are. At Didyma especially, he is not sure whether the diviner uses a wand, or sits on an axis, or makes some use of a sacred spring. In any case, the place, the actions, and the material all form an inseparable unity, which brings the oracle and the postulant into union with a higher being.

CHAPTER 3

DIONYSIUS THE AREOPAGITE

Vincent Scully, in his 1962 study of Greek temples in their landscape setting, argues that the ancient Greeks did not locate their temples without consideration to the surrounding landscape.[1] As a general statement, this point hardly needs arguing, since the temples appear in locations already hallowed by myth. The temple of Apollo at Claros, for example, is built over a sacred spring said to have originated in the tears of Manto, the daughter of Tiresias. It is difficult to find a sacred site in Greece whose sacred character does not depend on the action in that place of a mythological figure.[2] Scully, however, wishes to make a more specific claim: that features inherent in the landscape, rather than actions said to have been performed there, make the place holy and so a suitable site for a temple. Prominent among these landscape features are the horns of Cybele: two pronounced hilltops, which, from the temple site, appear close together. Scully recites the great number of temples whose landscapes exhibit this feature, as well as other temples, which possess other symbolic features. Such features serve to drive home Scully's claim that "the place is holy even before the temple is built."[3]

The temple, in Scully's presentation, does not simply reiterate in architectural terms the sacred character already present in the landscape. Greek temples "were so formed in themselves and so placed in relation to the landscape and to each other as to enhance, develop, complement, and sometimes even to contradict, the basic meaning that was felt in the land."[4] We have earlier seen Iamblichus describe this formation as the "working" of the place from its already sacred components. Walter Burkert, in his 1977 study *Greek Religion*, flatly denies this relationship between temple and landscape. He readily agrees that the Greeks mark off the land to be contained within the temple precincts, but he claims: "natural features are seldom appropriated for this purpose."[5] If natural features do play a role in identifying the sacred precinct, they do so only as particular things rather than as whole landscapes. For instance, the single sacred stone at Delphi serves to identify it as the "navel of the universe." The laurel grove at Didyma serves to identify it as a shrine of Apollo. Water sources also identify shrines, for practical as well as symbolic reasons, since the pilgrims must have a means of ritually washing themselves, and some form of sacred water is necessary for the sacrifices. Such natural features do not constitute the sacred precinct. For Burkert, the Greek sanctuary "is properly constituted only through the demarcation which sets it apart from the profane (*bebelon*)."[6] The boundary wall, or boundary markers, replace Scully's natural landscape features as the determinants of the temple site in Burkert's reading of Greek religious practice. Because the boundary wall now defines the sacred place, the distinction between sacred and secular is all the more marked. As Burkert understands it, the very term "sacred" (ἱερός) indicates its radical division from the secular: it has a "delimiting, defining function."[7] Such an understanding of the sacred goes directly against Eliade and Heidegger in "The Origin of the Work of Art," since the sacred in Burkert's sense "does not constitute the entire world and does not lay infinite claims on men."[8] The sacred has its own precinct, and it does not carry over into the day-to-day lives of those who worship in that precinct.

Our study of the Hellenic Neoplatonists reveals them to be in greater harmony with Scully's reading of Greek practice than Burkert's. Not only statues and temples have a divine allotment, but cities and whole regions as well. The sacred objects used within the temple precincts have a sacred character that overlaps with their day-to-day use. A dove, for instance, has a sympathetic connection with the gods whether it is sacrificed in the temple precinct or perched in a tree on an Attic hillside. While late Neoplatonists like Iamblichus do privilege specific sacred places, these places are sacred not because they differ from the rest of the world, but because they show a special affinity to one of the sources of that world. It would be only natural, then, that "the landscape and the temples together form the architectural whole."[9]

While Burkert's reading of the sacred does not square with the Hellenic Neoplatonic reading, he does unintentionally capture the early Byzantine Christian approach to the sacred, both as revealed in period architecture and as described in the work of the mysterious fifth- or sixth-century author known as Dionysius the Areopagite. The five works now known as the Dionysian corpus appeared in the early sixth century, each claiming to have been written by Dionysius the Areopagite, the Athenian convert of the apostle Paul mentioned in Acts 17:34. Even at the time of their appearance in the sixth century their attribution to Paul's convert was considered suspicious enough to require defending. By the early Renaissance, they were finally dated to the late fifth or early sixth century, and more recently they have been found to have probably originated in Syria. Their author, whoever he was, had enough firsthand knowledge of Hellenic Neoplatonism to quote Plotinus and Proclus, and to make use of ideas developed by his fellow Syrian, Iamblichus. One of the works within the Dionysian corpus, *On the Ecclesiastical Hierarchy*, contains the first thorough exposition of the rites of the Byzantine church as they occur within its sacred precinct, and so provides us with a reflection on the sacred place that closely matches the architectural transformation of the temple in Byzantine Christianity.

The Christian temple differs most radically from the Hellenic temple in its relocation of the altar from outside the main entrance to the most protected part of the interior. Because the Greek altar stands outside the temple, the exterior of the temple plays an important role in the sacrificial rites. Much of the activity in the sacred precincts takes place on approaches to the temple and makes use of its exterior features. In these cases, the temple "may become a fully sculptural entity. . .expressing its god by its own sculptural qualities: so making his character, otherwise hid, externally visible."[10] We have already seen how Iamblichus takes this to be the nature of the temple, to such a degree that he speaks of sacred places and sculptures in the same breath, as belonging to the same order. Burkert downplays the significance of the temple in the rites: "for the living cult they were and remained more a side-show than a centre."[11] For Burkert, the center of the cult activities is not the temple but the altar. Once the Christians move the altar indoors, the exterior of the temple yields pride of place to its ability to enclose, a twofold operation which lends unity to those within and protection from those without. This capacity to enclose, once one among many functions of the temple building, now increases so much in importance that, in Dionysius, it eclipses nearly all the others.

Once the interior life of the temple replaces its exterior life, we can see how the exterior orientation of the temple may become less important, a change of emphasis borne out by the archeological evidence from the fifth century. Churches from the earlier, Constantinian era, often orient themselves toward surrounding topographical features, but in the fifth century the outside landscape ceases to have any relation to the layout of the temple. As Richard Krautheimer puts it: "orientation becomes the rule. The church axis runs west and east: the apse facing east, the facade west—contrary to Constantinian church plans where the apse faced west, east, or north, subject to regional custom and local topographical conditions."[12] The church plan now has a "comparative homogeneity," which contrasts markedly with the varied character of the older Hellenic temples. The older temple takes on a specific form, according to Scully, "both from its adjustment to its particular place and from its intention to personify the character of the deity which it, too, is imaging there."[13] Scully's formulation recalls Vitruvius, who says: "temples should not be made according to the same principles for every god, because each has its own particular procedure for sacred rituals."[14] We will find in Dionysius an account of the rites that conforms quite well to the new architectural approach to the church building. For Dionysius, the sacred place is not a sculpture. It does not embody the god. Instead, it opens up a space in which the various kinds of human being—from the most absorbed in the body to the most intent on the intelligible—can enter into communion with one another and with the divine. This communion has an important spatial dimension. Those absorbed in the bodily cannot unite with their intellectual counterparts without a space capable of uniting them. Likewise, they cannot enter into communion with the divine unless they can orient themselves physically toward an object of contemplation. Dionysius never says, however, that they contemplate the sacred place itself. The altar, and not the sacred precinct, is the focus of contemplation.

The Christian Neoplatonic Landscape

Before we look at Dionysius, who concerns himself almost exclusively with the interior space of the church, we would do well to mention the ongoing interest among Christians in the landscape outside the church. Descriptions of actual landscapes are as rare among the Christians as among the Hellenic Neoplatonists. Landscape description inhabits the margins of Christian discourse: letters, biographies, and similes. Nevertheless, we see in these texts an explicit development of the Plotinian theory of stillness, an important and ongoing theme, however marginally expressed. D.S. Wallace-Hadrill's 1968 work *The Greek Patristic View of Nature* provides a general survey of passages where the Greek Christians describe the activities of visible living things, lifeless things, and the rural way of life that has to pay careful attention to them. Wallace-Hadrill does not emphasize the appreciation of landscape in particular—what he calls "scenery"—and he does not deal with Augustine, but otherwise his book lays the groundwork for what follows here.[15]

The Charming Landscape

Basil, the fourth-century bishop of Caesarea, has left us with one of the strongest early Christian evocations of a landscape outside the context of allegory. Having visited the monasteries of Egypt and the Middle East, he sought out a place where he might found his own monastic community. He reached such a place in Pontus, and describes his surroundings in a letter written in the year 360 to his friend Gregory of Nazianzus. He lives beneath a mountain, on a ridge of land extending down from the mountain, bordered on both sides by steep ravines. At the bottom of the ridge, where it flattens out, a river runs from ravine to ravine, effectively isolating the ridge on all four sides—river at the bottom, mountain at the top, and ravines on both sides. While Basil also describes the forest and wildlife of the region, this characteristic of isolation touches him the most. He says: "the greatest thing I have to say of the place is that it nourishes stillness, the sweetest of all fruits to me, though it is suited to bear all fruits on account of its felicitous position."[16] Perhaps because of this mention of stillness, Hilary Armstrong claimed a direct dependence on Plotinus here.[17] Their common use of this single term, however, does not seem to require that we assume a direct dependence, and doing so might lead us to overlook an important difference between the two. Plotinus invites his readers to imagine the stillness of the four regions of the earth as an exercise in stilling the mind, so that the pouring of soul into the world may be that much more apparent. Basil, on the other hand, has sought out an *exterior* landscape as an aid to the cultivation of his soul, much as Iamblichus' Pythagoreans sought out the stillness of the temple precincts so as to have a place well suited to the composition of the mind.

Where Basil differs from Iamblichus' Pythagoreans, and even more radically from Plotinus, is in his choice of uncultivated land as a place for the cultivation of stillness. We may infer the lack of cultivation here from Gregory's response to Basil: "whatever has escaped the rocks is a ravine, whatever has escaped the ravines is brambles, and whatever is above the brambles is a cliff."[18] And in a later letter: "how will I pass over those unfruitful and gardenless gardens?"[19] As Wallace-Hadrill has noted, Gregory's comments reveal that Basil has found stillness in a landscape substantially unmodified by human craft.[20] Gregory's mock horror reveals how unusual it would be for someone to treat such a landscape as still in Late Antiquity.

The dialectic between a stillness to be found in cultivated and uncultivated landscapes continues throughout the Byzantine tradition. Alice-Mary Talbot argues that both suitability for cultivation and the very lack of such suitability influences the choice of monastery sites in the medieval Byzantine tradition.[21] In the Life of St. Luke of Steris, stillness goes hand in hand with suitability for cultivation: "see what sort of place this is where you are standing—how temperate in climate, how pleasant, free from all disturbance and isolated from men," but also "supplied with very pure water, sufficient both for the demands of thirst and for the irrigation of vegetables and plants."[22] In the Life of Lazaros of Mount Galesion, on the other hand, we hear of a place that is "inaccessible, rocky, quite rugged, and waterless, too, and able for these reasons to provide much stillness to someone arriving there."[23] The tradition of praising monastery and garden sites because of their stillness continues throughout the Byzantine Middle Ages.[24]

If Armstrong's claim that Plotinus' visualization exercise had a direct influence on Basil's description of his hermitage is unjustified, he is on much firmer ground when he claims that Plotinus directly influenced Augustine's description of his vision at Ostia with his mother. They are standing, leaning against a window that looks onto the interior garden of the house where they are staying. The scene, then, already suggests the tranquillity of a walled-in garden. They have been talking about the eternal, and "while we spoke and gaped at it, we barely touched it with the whole force of our hearts, and we sighed and left there the first fruits of the spirit, and we returned to the sound of our mouths, where the word has a beginning and an ending."[25] Augustine and his mother wonder what would happen "if the tumult of the flesh should be silent for anyone, if the images of the earth and sea and air should be silent, if the poles should be silent, and if the soul herself should silence herself and transcend herself by not reasoning." Their answer is that God would speak in his own voice and not theirs, and allow them only the vision of him. "He would rapture, devour, and conceal his audience in interior joys, so that their life might be forever such as this moment of understanding had been." Augustine here stands within the Plotinian tradition, in which contact with the divine may be stimulated by characteristics within the sensuous world, but the contact is resolutely interior.

Augustine's positioning of landscape as a stepping-stone to a thoroughly interior experience of the divine may be Plotinian, but it does not lead Augustine to disregard the visible church in the way that Plotinus disregards the shrines of the Hellenes. Augustine does not say of God: "he should come to me, not I to him." Yet Augustine does not take up a philosophical treatment of the liturgy of the church and its unfolding in place. He does mention place here and there within his works and he does develop a sacramental theology, but he does not anticipate Dionysius by reflecting on the sacred place or setting his sacramental theology in an explicitly spatial context. Instead, Augustine shows regard for the church in the way that Plotinus shows regard for his school. Plotinus and, late in life, his student Porphyry, reject the shrines of the Hellenes because they misleadingly suggest that the gods enter into place. The school of Plotinus, however, does not meet with the same rejection. His school is the place of a philosophical community, where souls are developed in words, and communion between souls and the divine occurs.[26] In the same way, the church is the place of the believing community. Outside it, that community cannot find full expression. Unless we see this side of Augustine, we are at pains to account for the intense effect that the story of Victorinus has on him when he is yet a catechumen.[27] Augustine, not yet a member of the church, pours out his heart to Simplicianus, the future bishop of Milan, who tells him the story of how Victorinus, a professor of rhetoric like Augustine, became a Christian. When an old man, Victorinus carefully examined the scriptures and the writings of the Christians, then went to Simplicianus and said to him: "you should know that I am now a Christian." Simplicianus did not rejoice, but replied: "I will not believe it, nor will I think you a Christian until I see you in the church of Christ." Victorinus replied: "Do the walls, then, make us Christians?" Simplicianus recounts to Augustine that he and Victorinus repeated this exchange many times, until Victorinus came to him one day prepared to confess his faith within the church, having come finally to fear that his refusal to publicly affirm his Christianity might lead to Christ's refusal to acknowledge him in the age to come. The walls of the church here serve to divide the realm of private individuals from the communal body of Christians.[28]

The Sublime Landscape

We may take as typical of Late Antique attitudes toward uncultivated land Gregory of Nazianzus' response to the roaring river and forest thickets near his friend Basil's hermitage. But two near contemporaries, John Chrysostom and Gregory of Nyssa, both find the lack of measure in certain landscapes to be a source of wonder rather than disgust. Chrysostom mentions the ocean briefly as an object of wonder in his discussion of divine incomprehensibility. He identifies two species of wonder, one associated with what we would call the beautiful. This species concerns "the things we are content to wonder at, but without fear, like the beauty of columns, or the works of a painter, or bodies in their bloom." A second species of wonder involves fear directly, and should recall for us Damascius, who describes the effect of the waterfall as "august (σεμνόν) and terrible (φρικῶδες)," and as filling the spectator with "reverent fear (φόβου)." Chrysostom uses much the same language to describe our experience of the ocean. He says: "we wonder. . .at the extent of the sea and its unlimited abyss, but with fear (φόβου), when we look into its depth."[29] For both Damascius and Chrysostom, fear is the distinctive effect of this sort of landscape. What the majesty and power of the waterfall do for Damascius is accomplished for Chrysostom by the "infinite and yawning" depth of the sea. Unlike Damascius, Chrysostom does not use the term "terrible" (φρικῶδες) here, though it appears several times elsewhere in the work, and Jean Danielou has suggested that "the word essentially characterizes the [celestial] liturgy, or it expresses also the atmosphere of the earthly liturgy which is the participation of it."[30] Perhaps

Chrysostom prefers to restrict "terror" to liturgical rites, and not to natural phenomena. Such a distinction would, of course, have been meaningless to Damascius, whose tradition celebrates natural phenomena as places suitable for liturgical rites, once they have been "worked" by the priests, like the local custom at the waterfall he describes.

Gregory of Nyssa, concerned, like Chrysostom, with the infinity of God, chooses the example of a mountain peak as a visible image of this infinity. He invites us to consider someone on such a peak, who has "below him a smooth, vertically cut rock, which extends toward the base in an infinite depth, in a straight and polished form." The person in such a position will experience "what, in my opinion, the soul experiences which, passing by what is accessible to the concepts proper to extension, seeks out the nature which precedes time and which is not comprehended by extension. It has nothing by which it could be grasped: not place, not time, not measure, nor any other such thing which can receive the application of our reasoning."[31] Because of the limitations of our senses, vast distances appear infinite to our eyes, whether these distances be the depths of the ocean or the slope of a mountain. Their effect derives precisely from their appearance of having no place. They seem to go beyond extension. In this respect, they differ sharply from Damascius' experience of the waterfall. Damascius mentions nothing about lack of place. It is rather the power of the place that moves him. The gods, who admittedly appear there in a rather picturesque form, enter into the place that has the finite natural characteristics suited to them. It is this conception of place that Dionysius employs in his own work, though he restricts it to the interior of the church building, which acts to shape the liturgical rites in such a way that the divine may appear in them.

The Hierarchy of Souls in the Sacred Place

Like the later Hellenic Neoplatonists, Dionysius describes a rite that takes into account the inability of most human beings to engage their intellect directly. Such a rite makes use of sensible symbols to engage the intellect indirectly. The cause of this indirect engagement is the divine, which first of all supplies the symbols and then works through them in the course of the rites. Since the symbols are bodily, however, they also participate in the causal structure that belongs to bodies and to embodied souls, and this causal structure is not irrelevant to the rites. Iamblichus refers to the bodily contributions to the rites as "accompanying causes," ranked beneath the "preceding cause," which is the divine. The primary accompanying cause of the Hellenic tradition is the natural sympathy between bodies. The animal chosen for a particular sacrifice may depend on its sympathetic connection with the purpose of the sacrifice or the planetary god allotted to the place of the sacrifice. In the Christian rites, the sacred history described in the Hebrew and Christian scriptures takes the place of natural sympathy, and determines the choice of symbolic things to be contemplated in the course of the rites. Dionysius wraps his ritual theology around this sacred history, but he also treats as an accompanying cause, though he does not use the term, the bodily effect of the sacred place. The confines of the church building establish a rudimentary unity out of the many human individuals who gather there, and certain sensations like fragrance and sound reinforce this unity. Within this unity, the openness of the interior space allows the hierarch to unfold the intelligible content of the rites into a spatially divided multiplicity capable of being grasped by the laity. Though the divine is the primary cause of the rite, the sacred place acts as an accompanying cause, opening up a space in which the divine and human can interact. The degree to which a human being requires this place depends on the degree to which it can directly engage its own intellect. The laity, as we will see, need the sacred place for their salvation; the hierarch hands down that salvation through the sacred place.

The Body as Necessary for the Soul's Salvation

Like Iamblichus, Dionysius thinks of the soul as having two parts, one which can think without bodily influence, and one which requires bodily influence.[32] In his letter to Titus, Dionysius describes the soul as divided, according to its mode of knowing, into a passionless and a passionate part. Dionysius does not mean "passion" here in the usual sense of the lust for material things. This latter passion is the characteristic feature of the possessed, one of the classes of those outside the church, who "desire and engage in the change belonging to what is material and quite full of many passions."[33] The destruction of this passion is one of the features of baptism. Dionysius uses "passion" differently when referring to a continuing feature of the life of the faithful after baptism. They still exercise the senses, and so they are affected by things outside of them, even if those things no longer inspire lust. This meaning of "passion" is closer to its root meaning of "be affected." The senses cannot sense unless they are affected by outside objects, and so all knowledge gained through sensation is passionate when the term is used in this second way. "Passionless," on the other hand, means unaffected by outside objects, and so the soul's passionless part "is appointed for the simple and innermost visions of the deiform apparitions."[34] The passionless part of the soul sees the appearance of the godhead from within, in its innermost part, and needs no exterior aid to affect it from outside. The soul's other part, its passionate part, "extends itself to divine things by reshaping the figural symbols prepared for it beforehand, since such veils are appropriate for it. This is clear from the fact that even those who have heard theology distinctly, without its veils, nevertheless mold in themselves a kind of figure which leads them by the hand to the thought of the aforementioned theology." Dionysius himself often follows up a direct explanation of an intelligible truth with an example derived from the realm of sense experience. This "figure" or "veil" then aids his comprehension of the intelligible truth.[35] The passionate part, then, does not have its own object—it is not, for example, directed at the senses—but serves as a bridge between the sensuous and the intelligible, so that even those of us who need the sensuous may still have the knowledge that the soul's passionless part obtains directly.

When Plotinus identifies a higher part to the soul, he gives it such intellectual characteristics that it seems at times no different from the divine intellect. At first glance, Dionysius seems to face a similar problem with his passionless part of the soul, though here the object of comparison is not the divine intellect, but its Dionysian equivalent: the angelic intellect. Dionysius tells Timothy, bishop of Ephesus and the addressee of *On the Divine Names*, that the angelic minds "gather divine insight not in divided things, nor from divided things or from the senses or from definitions (λόγοι), nor do they encompass it from anything common to these."[36] Dionysius takes care here to insulate the angels in two ways from the divided character of our own human experience. First, angelic knowledge does not arise through the senses, and second, it does not arise from definitions, the discrete forms of things we find in the human soul. These definitions may be immaterial, but they remain multiple, since each definition is discrete from the rest, and so they join sensations as the "divided things" that the angels transcend. The angels, he says, "are purified from everything material and multiple, and think of the intelligible side of divine things intellectively, immaterially, and uniformly." Because their activity is undivided, immaterial, and unified, they are "synoptic (συνοπτική) of divine intellections." "Synoptic" refers to their ability to see many objects in a single glance.

Just how much do the angels see at once? Plotinus suggests that intellectual knowing will know everything in knowing any one thing, since all things are contained in each intelligible form. Though clearly an admirer of Plotinus, Dionysius never follows him on this point, and indeed, to do so would confuse his hierarchical system. Even the angels fall short of the divine mode of knowing, which is a pure unity. Human beings fall short of the angels since, even in their highest mode of knowing, they can only achieve "thoughts that are the equal of those belonging to the angels." Our souls typically think by moving from one discrete thought to another, but we also have the capacity to enfold those multiple thoughts into a unity. This does not make us into angels, since it necessarily changes only the mode of our thinking, not its source, which may remain the senses. By elevating the content of the senses, the soul can achieve a form of thought that is the equal of the angelic thought. Dionysius does, once in his entire corpus, say that we may "ascend, with otherworldly eyes, from unclear images to the author of all things, to contemplate all things in the cause of all things, uniformly and unitedly, even things opposed to one another."[37] We have no good reason, however, to assume that Dionysius means here that we will, outside of time and place, acquire synoptic knowledge of all things by knowing the divine being. He seems to mean simply that, when we see all things one by one, we should contemplate them in their cause, where they are present in a uniform manner. The form of the passage supports this interpretation, moving as it does in an ascent from unclear images, necessarily seen one by one, to their cause, rather than in a lateral motion from the cause to all things as seen in the cause. The passionless knowing of humans knows the causes of all things, but not the particulars of all things. Particular knowledge remains local, confined to what we encounter through the senses.

So far we have seen particular knowing in Dionysius only as the passionate knowing which begins in the senses and ends in the mind. Dionysius has a richer conception of particular knowing, which includes a form not dependent on the senses at all. He describes it with a geometrical metaphor. If the passionless mode of knowing is circular, and the passionate mode of knowing is a straight line, this new mode of knowing is a spiral, meaning it combines elements of both circularity and linearity. The metaphor suggests that this new mode of knowing constitutes the mean between the two extremes of passionate and passionless knowing. Is this the form of knowing possessed by Iamblichus' middle class, which is neither immersed in the senses like the great herd, nor free of the senses like the sage?

The soul experiences this form of knowing when "it is illuminated with divine insights properly to itself, not thoughtfully and unifiedly, but rationally and discursively, and as though with mixed and changing activities."[38] It is hard not to see in Dionysius' phrase "mixed activities" an echo of Iamblichus' middle class, who may engage in a mixed way of life, and a mixed rite. Yet Dionysius' middle mode of knowing cannot easily be reduced to that of Iamblichus. First of all, Dionysius has already combined sensation and thought in the linear mode of knowing (passionate knowing), which moves from sensation to the mind. Second, the middle way does not even mention sensation, and so does not seem to combine sensation and thought at all. We may wonder why Dionysius even uses the term "mixed," with its suggestion that this mode of knowing is the mean between two extremes. He gives us no clear answer, but the middle way does seem to be immaterial, like passionless knowing, but multiple, like passionate knowing. If this is the case, then Dionysius' middle way differs, markedly, from Iamblichus' middle way, since the multiplicity involved in the Dionysian middle way need not go beyond numerical multiplicity into sensuous multiplicity. The Iamblichan middle class, on the other hand, has the sensuous as part of its mixture. The Dionysian middle way of knowing is also "proper to us"—a claim Iamblichus never makes about his middle class. For Iamblichus, the middle class makes use of sensation, but perhaps owing to our only temporary involvement with the body in the Hellenic Neoplatonists, he wishes to avoid suggesting that this involvement with sensation is ever "proper" to us.

The historical characters with whom Dionysius populates his treatises, often treated merely as part of a ruse to gain credence as the first-century convert of St. Paul, actually serve to provide concrete examples of the different levels of human knowing. In this respect, Dionysius takes a page from Proclus, who creatively interprets historical figures from the history of philosophy, such as Parmenides and Zeno, as representative of different levels of knowing.[39] Pride of place in Dionysius' cast of characters must be given to the Apostle Paul and Hierotheus. Hierotheus, a probably fictional bishop whose discourses taught Dionysius "the elements,"[40] shares with Paul the ability to receive the divine by the effacement of the human. Dionysius says: "the great Paul, who became erotic in his possession by the divine, and transported with its self-effacing (ἐκστατικῆς) power, said with his divine tongue: 'I live, yet not I, but Christ lives in me.' "[41] Dionysius tells us that Paul speaks here "as a true lover and, as he himself says, effaced by God, living not his own life, but the life of his beloved."[42] We have already seen how the language of self-effacement appears in Plotinus and Iamblichus as a means of participating in a higher principle through the rejection of one's own nature. Here we see Paul living out this self-effacement, as Hierotheus also does. Dionysius tells us that Hierotheus wrote a book entitled *The Elements of Theology*, filled with "whatever was handed down to him by the sacred theologians, whatever he synthesized by a scientific search of the scriptures with much exercise and labor over them, and whatever he was initiated in by a certain more divine inspiration."[43] The first two sources are sensuous and discursive; the last is not. Dionysius describes this last source as "not only learning but suffering divine things, from a sympathy with them—if we have to put it this way—perfected by a mystical union and trust in them that is untaught."[44] Dionysius here includes a Neoplatonic technical term, "sympathy," with the usual disclaimer that he attaches to such terms: "if we have to put it this way." He does not use "sympathy" in its technical sense of an occult bond between things, but as a complement to his claim that Hierotheus is "suffering" divine things. Both "sympathy" (συμπάθεια) and "suffering" (παθών) rely on the root *path-*, which indicates receptivity or passivity. Hierotheus becomes the passive receptacle of a higher power, just as Paul does. This method of "suffering divine things" is the one by which the apostles handed down the liturgy and the scriptures, receiving them intellectually—"from intellect into intellect"—from the angels. The same method is employed by the hierarch during the liturgy. While "the many look to the divine symbols only, the hierarch is led up by the thearchic spirit to the holy principles of the rites in blessed and intelligible visions in the purity of his deiform disposition."[45] Like the Iamblichan sage, the Dionysian hierarch, personified by Paul and Hierotheus, has no need of the material to enable his contemplation of the intelligible, but he does deliver the fruit of his contemplation to the lower levels of the hierarchy. However, where the Iamblichan sage may deliver oracles on a regular basis, the apostles and theologians have handed down the scriptures and liturgy primarily in the past. The hierarch now emerges from contemplation in the course of the liturgy not to deliver oracles, but to cense the nave, making a "procession into the secondary" before his return "to the contemplation of the primary."[46] This censing, as we will see, brings symbols of the intelligible into the sensuous and multiple character of the nave, where the laity may contemplate them meaningfully.

What of the middle class? Dionysius frequently describes our usual mode of knowing as involving the senses. While we may, in the life to come, participate in "the intelligible gift of light in our passionless and immaterial mind," we presently have recourse to a different source and method of knowing. We "use fitting symbols for divine things, and we are elevated proportionately from these again to the simple and unified truth of the intelligible visions." When Dionysius says "we" here, he includes himself, even though he is a hierarch. Timothy, his addressee, is also a hierarch, but should also be included among those that need the aid of the sensible in contemplating the divine. Dionysius reminds Timothy of the time when Timothy returned the *Elements of Theology* to Dionysius as being "too lofty" for his understanding. Dionysius responds that he himself is lesser than Hierotheus. He says: "to see the intelligible discourses and their synoptic teaching with one's own eyes requires the power of an elder, while the knowledge and learning of the structures that lead to this is fitting for lesser initiators and their charges."[47] He often describes Hierotheus as an elder, and here he reminds us that the power of an elder is to see the intelligible directly (meaning without reference to the sensuous). Dionysius himself is only one of the lesser initiators, who hopes "to unravel and to separate by a reason proportioned to us the synoptic and unitary enfoldings of that man's great intellectual power."[48] We would be wrong to try to identify Dionysius' own way of knowing as either the spiral or the straight motion mentioned in his geometrical metaphor. He uses both, since both involve multiplying out what is in itself a unity. The spiral motion does so through discursion, which is not sensible, but involves a numerical multiplicity. The straight motion does so through the sensuous, which spatially distinguishes what is in itself a unity. Nor should we separate Dionysius' mode of knowing from that of the laity as the Iamblichan middle class is separated from the great herd. Dionysius never implies that the laity are restricted to sensuous action. They move from the sensuous to the intelligible just as Dionysius does.

The ranks of the hierarchy which stand below the laity—the penitents, the possessed, and the catechumens—come the closest to filling the shoes of the Iamblichan great herd. The Dionysian rites reveal sacred things, which must be contemplated symbolically, as embodying intelligible realities. The penitents, possessed, and catechumens are removed from the church before the sacred things are brought into view because they have no familiarity with the intelligible, and so cannot yet contemplate such things symbolically. The possessed, as we have already seen, have as their distinctive characteristic the immersion in "what is material and quite full of many passions." Dionysius does not say much about the penitents as a distinct class, but we may assume that they have acted in a way that misunderstands the relation of material and intelligible, and this has led to their present status. The catechumens may not be immersed in the material, but neither are they suited yet for the ritual elevation to the intelligible through the material, since they have not been baptized. The characteristics which, for Dionysius, are *prima facie* reasons for excluding these classes from the rites are, for Iamblichus, the very reasons for bringing them into the rites: they live at the level of body. But then, as we have seen, the Iamblichan rites allow the sensible to participate in the intelligible through action, regardless of whether the participant is able to interpret the sacred things symbolically.

No ecclesiastical rite aids the penitents, the possessed, and the catechumens, but the hierarchy does not neglect them completely. The deacons have as their special task the education of the uninitiated, and Dionysius himself provides a summary course on what the uninitiated ought to be told about the rites. He begins his discussion of baptism, for instance, "keeping silent about the more divine structure of the rites," and providing "an introductory guidance for the uninitiated."[49] The more divine structure of the rites is their intelligible structure, which is suited only for initiates, since it is embodied in the sacred things themselves. The "less divine structure" of the rites, so to speak, is the ethical allegory they reveal. Dionysius tells the uninitiated that the baptismal immersion of the initiate in water teaches that we should purify ourselves from evil. When Dionysius interprets baptism for the initiated, he will set aside this ethical instruction in favor of the intelligible structure embodied by the rite: the structure of division and communion.

The rites, then, are properly for the middle class—the laity—which is able to exercise the passionless part of the soul only in combination with its passionate part. For such a class, the rites are not an ethical allegory of purification, but a "road" and a "leading by the hand" to the intelligible.[50]

Is There a Divine Allotment?

The laity need the sacred place in which the hierarch may present symbols that, when contemplated, can elevate them to the divine. So far, we have not seen any evidence that Dionysius follows the Hellenic Neoplatonists in taking the further step of recognizing specific things as sacred to specific places for specific people, under the providential care of specific gods. The laurel tree at Didyma, for instance, was sacred to Apollo, and was especially significant to people under the protection of Apollo, and to citizens of cities whose special guardian was Apollo. The veneration of specific places, things, and deities did not die out when the communities of the Mediterranean converted to Christianity. Patron saints took the place of guardian deities, both for people, cities, and shrines. Dionysius himself says surprisingly little about the ties between specific saints or angels and specific people, churches, cities, and regions. He does tell us that the hierarchy of living human beings within each church community is headed by a single, specific hierarch: their living bishop. Since there are many different churches, there will be many different hierarchs, and so we may speak of multiple human hierarchies.[51] He says nothing about specific churches being under the protection of a patron saint, but he does advise individuals to seek the mediation of specific living saints: "I say, following the scriptures, that prayers of the saints in this life are altogether useful: if someone desires the sacred gifts and has a sacred disposition to participate in them, as one mindful of his own shortcoming, and goes to one of these holy men, and asks him to become his partner and fellow suppliant."[52] The assistance of these living saints simply follows the pattern laid down in the formal hierarchy of the church: "the divine gifts should be given to those worthy of participating in their divine rank through those who are worthy of transmitting them." Dionysius here speaks not of departed saints, but of the saints who are alive and capable of interacting with us visibly in the rites of the church. Those who think themselves capable of "fellowship with the godhead and who look down on the holy men" are mistaken not because we require the mediation of the departed saints, but because we need the mediation of the living. The laity, at least, need to see and interact with the human beings they strive to resemble. Simple commitment to an invisible being, or to the memory of the dead, is not a meaningful option for them. Such an option would take them away from the living hierarchy that surrounds them into a form of private piety, not capable of providing communion with God and the ecclesiastical hierarchy.

The angels, on the other hand, cannot be differentiated into the living and the dead. In what way do the angels serve as mediators? Dionysius never mentions the guardian angel, which he could have interpreted with reference to either the Judaic or the Hellenic Neoplatonic tradition.[53] The absence of such a reference, especially in an author so concerned with the angelic hierarchy, again suggests that Dionysius wishes to emphasize the importance of the ecclesiastical hierarchy against the forces growing in his own tradition that seek to circumvent it through various forms of personal piety. Dionysius makes one mention only of a divine allotment such as that described by Iamblichus. It is an allotment of the angelic powers, not of God, and pertains only to whole peoples, not to churches or cities. Dionysius carefully rules out the possibility that God, like Apollo, might be allotted only a certain nation, while other gods are allotted other nations. He says: "there is one principle and providence over all things, and we must not suppose that the godhead rules the Jews as its inheritance, while certain other angels or gods, of the same rank or opponents, had equal charge over other peoples."[54] The godhead presides over all peoples, but it illuminates them through the mediation of different angels. The angel Michael is the angel of the Jewish people; the Babylonians and Egyptians have their presiding angels; even Dionysius' own people—the Greeks—have their presiding angel. The angel can do its job even if the ecclesiastical hierarchy has not been formally established among the people, as is the case for Dionysius' Greeks and in the Hebrew scriptures, Melchizedek, who is not an Israelite and yet is described by the scriptures as "the priest of the most high God."[55] Dionysius cautions us that Melchizedek is not a servant of "strange gods." There is "one principle of all things, and the angels who are the hierarchs of each people lead those who follow them to it." Melchizedek follows them in such a way as to form part of a *de facto* ecclesiastical hierarchy. He is called a priest, Dionysius says, because "he was not only converted to the truly real God, but in addition, as a hierarch, he led others in their elevation to the true and only godhead."[56] In Melchizedek we find, in a people neither Jewish nor Christian, a functioning ecclesiastical hierarchy owing to the allotment of an angel to each nation. Though Dionysius describes the angelic allotment in terms familiar to a later Neoplatonist, he here relies on a Judaic tradition, present within the Hebrew scriptures, and passed down to the Christians through Origen.[57]

The illumination of the angels radiates through the ecclesiastical hierarchy in two ways, one of which is inseparable from the sacred place, while the other, though independent of the place, finds its proper context there. These are the two traditions of the liturgy and the scriptures. Both come to us through the angels, who themselves have received the traditions "intelligibly and from within," enlightened with a ray that is "pure and immaterial."[58] What these angels receive immaterially from within themselves, they pass on to the ecclesiastical hierarchy in a form that is suited to the human soul, that is, "in the diversity and multiplicity of divided symbols." The two sets of symbols—the scripture and liturgy—are in different ways the verbal expression of the mythic context that allows the elevation of the soul. Since the liturgy is not written down, it is less material than the scriptures, which then occupy a lower rung on the scale of reality. The initiation of the liturgy is handed down immaterially from the angels to "our leaders" as "from intellect into intellect," and they have revealed it to us "through the medium of words that are bodily, yet at the same time immaterial, since they are free from writing."[59] The liturgy serves as a kind of mean between the immaterial initiation of the hierarchs and the material initiation of the scriptures. The scriptures themselves may be differentiated using the similar standard of transparency (less material) and opacity (more material). The epistles and gospel (Christian scriptures) are clearer, frequently describing intelligible truths directly, while the psalms (Hebrew scriptures) are more opaque, using sensuous imagery.[60] On the other hand, the scriptures do not suffer by comparison with the liturgy, since "the writings handed down by God are the substance of our hierarchy."[61] The sacred history that constitutes the scriptures yields the symbols that the liturgy employs. Though the liturgy expresses the context in which the symbols can become meaningful, the text remains the substance of the symbols. When the text is brought into the liturgical context, the divine appears, indirectly, in the verbal and visible symbols employed in the rite. This confluence of scripture and liturgy in the sacred acts of the rite requires a place in which it can come about. The "diversity," "multiplicity," "division," and bodily character of both sets of symbols are not forms of numerical multiplicity, which rarely interests Dionysius, but forms of extension, distinction, and separation within a single place.

Vision and Pervasion

The Christians by the time of Dionysius refer to their eucharistic rite as a "gathering" (σύναξις), apparently deriving the name from the name for the Jewish sacred place: the "synagogue" (συναγωγή) or "gathering." Dionysius himself suggests that it ought to apply not only to the Christian eucharistic rite, but to all the Christian rites. He says that it is "extended even to the other hierarchic rites, though to this one selectly above the rest. It is given the common name of 'communion' and 'gathering,' since every sacral event gathers our partitioned lives into a one-like deification, and gives us union and communion with the One by means of a deiform enfolding of our divisions."[62] Every rite is a gathering because every rite overcomes our divisions in favor of "union and communion" with the One. The gathering has intelligible and visible components; it is a gathering of both soul and body. The soul achieves communion "in the pure contemplation and knowledge of the rites."[63] The body also achieves communion in the rite, not with the intelligible content, but with the symbols that manifest the content in the visible. The communion is brought about through visible sacred things such as "the most divine ointment. . .and the most sacred symbol of communion in the godhead," meaning the bread and wine which have become the symbols of Christ. The soul and body do not occupy different worlds. They perform the rites as part of the "whole human being," and the divine custom which established the rites then "sanctifies the whole human being, and sacredly works its whole preservation."[64] The communion goes further than a union between the divine and one human soul and body. It brings about a union between the various members of the human community and between the various parts of a single person. When Dionysius says that communion "gathers our partitioned lives into a one-like deification," he could be speaking of either the divided life of a single person or the lives of several people. Both types of partition must be overcome in the course of overcoming the further partition of our lives from the divine. As René Roques puts it, "union with God and union with oneself does not occur without union with the Christian community which is itself united to God."[65] Because the members of the ecclesiastical hierarchy have bodies, and for the most part experience the world only through these bodies, the rites do not simply involve "spiritual union" of souls with the divine, but also "spatial reconciliation" within the bodily confines of the church.[66]

The nave of the church perhaps appears more suitable to us as the place of gathering than it once did to scholars of early Byzantine church architecture. The remains of early basilicas in Constantinople, for instance, indicate that the church building was divided into several different compartments, each possibly meant for different orders. The altar area, side aisles set off by a colonnade, and upper level galleries all surrounded and opened onto the nave of the church. Richard Krautheimer has suggested that, by the fifth century, only the clergy entered the nave during the liturgy, while the laity confined themselves to the side aisles and galleries.[67] Even if some of the laity were allowed into the nave, women have been thought to be restricted to the aisles and the upper galleries during the liturgy. If the laity, or even solely female laity, were excluded from the nave, the unity Dionysius speaks of as the spatial accomplishment of the liturgically experienced church would be compromised. In his 1971 study of early churches in Constantinople, Thomas Mathews analyzes the historical texts on the liturgical uses of these churches, and concludes that the laity were not at this time excluded from the nave.[68] Male and female laity did likely occupy different sides of the central aisle, but they found themselves within the same space. As we will shortly see, the central aisle serves as part of a larger liturgical function of breaking up the amorphous unity of the space in favor of an orientation toward the altar. The rites do not simply serve to unite the members of the church, but to hold this unity in tension with a bodily orientation toward the altar.

Pervasion

Any time a group of people gather in a single place, they become one. Their bodies remain distinct, they may disagree with one another, but their place, by surrounding them, gathers them together and unites them in this restricted sense. Sometimes this unity is nearly adventitious: a group of people standing under a bus shelter in the rain does not participate in a particularly meaningful degree of unity. Even this low level of unity, however, can prompt communication within the group based on the shared experience, the getting out of the rain, that their gathering represents.

The "sacred order" of the church also becomes one simply by virtue of being united in the sacred place: the church building. In this case, however, the members of the group share more than a chance experience. Dionysius notes that the gathering does not result from the mutual actions of the laity. They owe their presence to their hierarch, who "gathers the entire sacred order into the sacred precinct."[69] By locating the immediate cause of the gathering in the hierarch (though even he receives it through higher channels: the apostles, the angels, and finally the godhead itself),[70] Dionysius removes the laity from responsibility for the rite. If the laity created the rite that they perform for their own preservation, they would be in the untenable position of raising themselves up by their own bootstraps. The rite must reflect the character of the divine and not the human, the cause and not the effect, and so it must be given by the cause, whose representative in the liturgy is the hierarch.

Dionysius says that the hierarch "gathers the sacred order" into the church in the context of preparing for the rite of baptism.[71] In the same passage, Dionysius rhythmically repeats with other verbs the prefix used in the Greek word for "gathering." The prefix is *sun-*, meaning "shared-." The hierarch "gathers" (συναγαγών) the community into the church for "shared work" (συνεργία), and "shared celebration" (συνεορτάσις) of the baptized person's salvation. The force of the prefix, which, as we have seen, characterizes not only actions within each rite but the goal of every rite—each one is a "gathering" (*σύναξις*)—emphasizes that the rite occurs only in tandem with the union of the participants in the place. The "work" to be done is a "shared work."

Union in a place is embodied and reinforced through the use of senses other than the sense of sight. The sense of sight sets up a spatial opposition between the one who sees and the object that is seen. I know that anything I can see outside my body is different from me, simply because my sense of sight spatially distinguishes it from me. Dionysius wishes to privilege the sense of sight for the vision of the divine things revealed on the altar: the ointment, bread, and wine. When viewing the divine things, the difference established by sight between the human perceiver and the perceived symbol is helpful, because it orients the perceiver toward the symbol without identifying the perceiver with the symbol. When it comes to establishing communion, more important is that there be some identification of perceiver and perceived, and so more useful for this are the senses that do not set up an opposition between the perceiver and perceived. Chief among these senses, at least in regard to the number of times Dionysius mentions it, is the sense of smell. Unlike light, which becomes less sensible as it passes through a space, a smell can *only* be sensed when it passes through a space. Smells distinguish space by pervading it, just as light distinguishes objects by illuminating them. This similarity between light and odor does not go unnoticed by Dionysius.[72] He uses parallel language to describe them. In discussing the fragrant oil that is consecrated for use in the various rites, he says, "those who participate in its fragrant quality do so in proportion to the amount of their participation in the fragrance."[73] This seemingly tautological claim actually means to say only that not everyone participates in the fragrance to the same degree. He goes on to explain the most basic condition of this participation: "the perception of sensible fragrances makes our faculty of smell feel good and nourishes it with much pleasure, if it is undamaged and oriented toward the fragrance."[74] We can only enjoy a fragrance if our sense of smell is functional and in a place where it can be reached by the fragrance. Dionysius uses the same language when speaking of light. He says both that we participate in light in proportion to our capacity to receive it, and that "light acts on things that are able to receive it."[75] Dionysius employs the sensible fragrance as he does sensible light: as a symbol for the intelligible appearance of the godhead in the human mind. But where he refers to the intelligible appearance of light as "illumination," he refers to the intelligible fragrance as a "pervasion" (διάδοσις).[76] The fragrance is not itself a determinate object, but takes the form of the space it pervades—in this case, the intelligible space of the human mind. Like light, which has no intrinsic shape but illuminates the space it fills, fragrance is an ideal symbol of divine communion, since the gift of the godhead to the receiver takes a form dependent on that receiver. In itself, it transcends form and cannot be comprehended.

Yet fragrance is not merely a symbol for Dionysius. Unlike light, its intensely sensuous character has its own role to play in the course of the rites. When the hierarch traverses the church with the censer swinging from his hand, the smell of the incense pervades the sacred place. The hierarch, Dionysius says, "goes from the divine altar all the way to the far ends of the shrine with the fragrance."[77] The entirety of the sacred place, then, participates in the same fragrance. Unlike the later vision of the things on the altar, where different orders of the hierarchy see the symbols from different distances, the smell of the fragrance is the same for each person in the sacred place. The fragrance does not merely symbolize a pervasion: it literally accomplishes a pervasion, which, through the senses of the laity, binds them together within the place. When Dionysius says that the action of the censing mirrors the rite as a whole, which "is multiplied out of love for humanity into a sacred diversity of symbols," and then "gathers them all into its own unity," he exhibits an understanding of multiplicity inseparable from place. It is not simply extension in number, but extension in mass as well. We experience the sacred symbols, in these cases, by their entry into place.

Sound, too, is capable of producing a community united in place. All of the rites of the church include the singing of hymns, but Dionysius mentions them explicitly in his discussion of the eucharistic rite. He tells us first: "the sacred writing of the divine odes has as its end the praising of all theology and theurgy and the extolling of the sacred words and sacred works of divine men. It produces a general ode and narration of divine things. It makes a suitable disposition (ἕξις) for the reception and transmission of every hierarchic rite in those who sing it with inspiration."[78] The hymns here play in their own way the same role as the singing of the psalms and the reading of the scriptures, which Dionysius describes just before he takes up the description of the hymns. All of them, by recounting the sacred history, and so making the rite's mythic context verbally present, prepare the listener to receive the divine form. In Platonic terms, they play the role of an accompanying cause, giving to bodies a disposition that is capable of receiving the divine form. Dionysius does not here focus on the fact that the hymns are sung, perhaps since the psalms and scripture readings are chanted, and so are themselves sung. He focuses instead on the power of the hymns to unite the faithful in the sacred place. He says "the hymnology, by means of the like voices (ὁμοφωνίᾳ) of the divine odes, establishes a like disposition (ὁμοφροσύνην) with the divine things, with themselves, and with each other, as though by one and like minded (ὁμολόγῳ) chorus of saints." Using one of his favorite methods of driving home a point, Dionysius builds up a rhythmic repetition of the prefix "like-" (ὁμο-) here. It appears three times, first to describe the fact that everyone in the sacred place is singing the same thing, so there is a likeness of voice among them; second, to describe the unity of disposition that results from this; third, to describe the fact that they look like they all think the same thing. The effect of the hymns here is not symbolic, but literal. The faithful and the clergy sing the same song, and this builds up in them a unity of disposition. They may not think the same things, but the fact that they now feel the same things crosses the gap between their different minds and makes them as though they were like-minded. This union can occur because of the pervasive character of sound, which allows a single person to add her voice to a chorus of voices, so much so that her voice is no longer separate from the chorus, but "homophonous" with it. After the singing of the hymns, the scriptures are read, and Dionysius leaves behind the pervasion imagery to take up again the language of symbol.

We may speak of one more of the five senses that brings about a union among the faithful who have gathered in the sacred place.[79] After the bread and wine have been brought out before the faithful in what is now known as the great entrance, "there is the holy working of the most divine kiss,"[80] later identified simply as "the working of peace,"[81] and known to us as the kiss of peace, in which the clergy and the faithful embrace each other. Dionysius explains why we associate peace with this embrace: "it is not possible to be gathered to the One and to participate in peaceful union with the One if we are divided against ourselves." Union with the One is peaceful, meaning it does not share in the division that divides us against ourselves. "Peace," in other words, is a term for unity, and unity is embodied by physical contact between the faithful. This contact is not primarily symbolic, since it *is* the unity it seeks to bring about. As with the singing of hymns, Dionysius speaks here of the action as an establishment rather than a symbol. The kiss of peace "establishes a uniform and undivided life, rooting like in like." The phrase "like in like" is ancient in origin, but among the Neoplatonists it is used specifically to describe the reciprocal likeness of one form to another.[82] One human being can reproduce another human being that is no more or less human than the first. This is why their likeness is reciprocal—the two human beings are exactly alike with regard to their humanity. Their likeness must be distinguished from the nonreciprocal likeness between lower and higher levels of form. In these cases, the lower form is like the higher form, but the higher form is not like the lower form. The portrait of a man is like the man, but the man is not like his portrait. Dionysius uses the phrase "like in like" to indicate that the communion brought about between the different people in the sacred place is a reciprocal form of likeness, and thus unlike their communion with God.

In the eucharistic rite, the uninitiated are removed from the church after the singing of the hymns and the proclamation of the scriptures. In the rite that consecrates the ointment, they are removed after the circuit of the censer around the holy place, and the singing of the hymns.[83] In both cases, the uninitiated participate primarily in those rituals that bring about a union of the community through the pervasion of fragrance and voice, but are removed from the church before those rituals that orient the faithful toward the contemplation of the symbols at the front of the church. Dionysius explains that they are removed precisely because they are unfit for contemplation: "they do not participate in the power of seeing the sacred things through the divine birth which is the giver and principle of light."[84] The light needed for contemplation is given by God in baptism, and until that light is given, the sight of the sacred things can only do harm. If we consider the actual practice of the church, this distinction is not quite as tidy as Dionysius makes it. The church is filled with symbols for the contemplation of the faithful from the moment they enter it. Likewise, the pervasion of sensations in the sacred place continues even after the catechumens leave, for example, in the kiss of peace. Dionysius has, however, made clear to the reader that the gathering of the faithful within the sacred place is engaged in two different activities, "spatial reconciliation" and "spiritual union." These activities may be more entangled than he allows, but they must not be reduced to one another.

Vision

The Iamblichan material rite serves as a means for the great herd to participate in the life of the gods, but their participation is resolutely material. The rite neither drags the gods down to the material realm nor elevates the herd to a contemplation of the intelligible reality, but it allows participation in the divine by the herd through the action of the rites. As we have seen, Dionysius does not allow any human character comparable to the Iamblichan great herd into the full performance of the rites. Only those human beings capable of both performing *and contemplating* the rites are initiated into them. What Dionysius has done is not quite what Paul Rorem claims in his 1984 study, *Biblical and Liturgical Symbols Within the Pseudo-Dionysian Synthesis:* "Dionysius transformed. . .a key emphasis in Iamblichean theurgical theory, for he stressed the anagogical value not of the rituals themselves but of their interpretation."[85] Rorem's claim here suggests that the value of the rites lies in reading them like a text, and understanding why certain actions are performed within them. If this were true, then the reader of Dionysius' *Ecclesiastical Hierarchy* would receive the same benefits as the lay person who goes into the sacred place and performs the liturgical rites. Rorem's underestimation of the necessity for ritual performance goes hand in hand with his later neglect of Dionysius' interest in the place of the rites, what Rorem calls "the Dionysian indifference to the church building."[86] Rorem is right to say that Dionysius emphasizes the necessity of interpretation in a way that Iamblichus does not, but this interpretation occurs primarily within the sacred place, and within the context of the liturgy. We see the new Dionysian emphasis in his changed description of the place in which the rites occur. Iamblichus says no more about the specific structure of the sacred place than that it makes space for the gods and that, as a whole, it serves as their symbol. Dionysius takes greater care in establishing the place as a means of unity between the individual initiates. He also articulates the various ways in which the place and the rites within it orient the initiates toward contemplation, the vision of the symbols which yield the intelligible reality.

The interior of the church building by Dionysius' time possesses an axial orientation in relation to the motion of the heavens. The main doors of the church line up along an east/west axis, with a central aisle extending along this axis from the royal doors, those that connect the narthex to the nave, through the inner sanctuary doors to the altar. Thomas Mathews has suggested that this "longitudinal emphasis" is connected with the entrance procession of the celebrants into the church, and that the continuing use of the basilica plan in Byzantine architecture reflects the importance of this procession.[87] In the time of Dionysius, the liturgy begins with the procession of the clergy into the church with the gospel, followed by the faithful. Dionysius himself does not make a great deal of the longitudinal emphasis, but he does set up the longitudinal structure in his discussion of the baptismal rite. The baptismal candidate is stripped naked and made to face west, the significance of which is only intimated by Dionysius as "unlit communion with evil." After the candidate renounces evil and spits in a westward direction, he is turned "toward the dawn"—toward the altar, though Dionysius does not mention this explicitly—and informed "he will stand in the divine light."[88] Dionysius does not make much of the celestial grounding of the longitudinal axis, preferring to stress the various motions that orient the participants toward the altar during the course of the rites. He addresses three orienting motions over the course of his work: the censing of the church, the elevation of the ointment, bread, and wine at the altar, and the relative location of the consecrated orders within the church building.

We have already seen that the incense, by pervading the church, plays a unifying role. The motion of the hierarch as he censes, however, plays an orienting role, which stands in a productive tension with the pervasion of the incense he spreads. The hierarch "goes with the fragrance from the divine altar to the far ends of the shrine, and is finally restored again to the altar."[89] Many scholars have noted that the motion of the hierarch here literally processes from and returns to the altar, and so symbolizes the procession and return structure of the Neoplatonic cosmos: each level of the cosmos proceeds from its prior, and returns to its prior. The intellect proceeds from the One; the soul proceeds from the intellect. What has not always been noticed is the nature and extent of the symbolism here. Dionysius does not draw our attention to the procession and return structure of the cosmos. The hierarch's motion symbolizes salvific rather than creative activities: his own activity as hierarch, the activity of God in the rite, the activity of the rite, and the activity of Christ. The symbol of the censer here is overloaded with meanings, an unusual move for Dionysius, who usually holds to a one-to-one correspondence of symbol to intelligible reality. The literal act of procession from and return to the altar serves to orient everyone in the church toward the altar, with the concomitant orientation toward these multiple symbolic contents the altar bears.

The hierarch's path through the church is first of all a symbol of his total activity as a hierarch. The hierarch, "in the form of the good, hands down to his subordinates his unitary knowledge of the hierarchy, using a multitude of sacred adumbrations." The hierarch breaks up his unitary knowledge, which is intelligible from beginning to end, into a multiplicity of symbols appropriate for leading the laity to this knowledge. The procession with the censer is simply one example of this. In this case, the hierarch generates a multiplicity by his very motion around the interior of the church. He does not himself become absorbed in this multiplicity, but returns to the altar, where "he is restored again to his own principle, unstained and unrelated to the lesser, and makes his own intellectual entry into unity." At the altar, the hierarch literally abandons the sensation and action of his procession around the church, and returns to the intellectual exercise of his mind. At the altar, he does not imagine other sensible things, other places, and times. These are not intelligible, and would separate him from the rites. The hierarch "sees, in purity, the uniform structures of the rites," that is, their intelligible principles. His intelligible contemplation is literally the source and goal of the rites, since he moves from this contemplation to the performance of action, with the end of leading all to participate indirectly in this contemplation.

The procession with the censer also symbolizes the entry of the godhead into communion with the sacred assembly of the church. Dionysius explains: "the blessed godhead proceeds by divine goodness into communion with its sacred participants, but it does not come to be outside the motionless standing and ground of its substance." Though Dionysius treats this procession as parallel to the procession of the hierarch, there are important differences. The hierarch may be "unstained" by his procession into the multiplicity of the nave, but he does move. The godhead, on the other hand, "proceeds into communion with the saints who participate it, but does not come to be outside the motionless stasis and ground of its substance."[90] When the godhead enters into the bread and wine, so as to bring about communion with the faithful who receive them, it remains motionless in its substance. When the gifts are elevated above the altar, the godhead does not rise. If the godhead were susceptible to such manipulation, the cause would be at the mercy of its effect. The real point of comparison between the hierarch and the godhead in the procession through the church is that they both begin at the altar, which is characterized not by motion but by its absence. This becomes ever more true as, over the centuries, the chancel screen grows to block ever more of the movement of the clergy within the altar area. The hierarch begins and ends at the motionless altar; the godhead always remains in it, even when it gives itself in communion. For this reason perhaps, Dionysius describes the altar as the "more divine place" within the sacred place that is the church building.[91]

A motionless source also grounds the other two symbolic meanings of the censing procession: the liturgy itself and the incarnation of Christ. As the censing procession multiplies itself, so too does the liturgy, by becoming actions and words distinct in place and time. Dionysius says: "the divine rite of the gathering—even if it has a unified, simple, and enfolded principle—is multiplied out of love for humanity into a sacred diversity of symbols and makes space (χωρῆ) for all the hierarchic iconography."[92] The rite has a "unified, simple, and enfolded principle," its intelligible content, which the laity contemplate through the various symbols it employs. The rite does not remain in the diversity of its symbols, but "is gathered again from these things into its proper monad, and makes one those who are sacredly raised up to it." The actions and words of the rite serve to orient the initiate toward its motionless and intelligible ground. Since, as Dionysius says elsewhere, the rite is nothing other than an image of Christ's incarnation, we are not surprised to learn that Christ's incarnation, too, unfolds from a simple principle into multiplicity, and returns again to unity. Christ's incarnation "gave communion with it to us as though we were of its kin."[93] Though Christ made himself one with us, the love by which he did so retained "the disposition of its own properties unconfused and undisgraced."[94] Christ does not change by his entry into communion with us. It is rather we who are changed, since the incarnation "showed us, so far as possible, an otherworldly elevation and a divine republic in our sacred likeness to it."[95] Dionysius immediately says that the means of this elevation, though revealed by the incarnation, is accomplished in the rite: the "sacred words" and "sacred acts," which remind us of God's sacred works in the incarnation itself and become the medium for continuing communion with it. The structure of the incarnation emerges from and returns to a unity; the structure of the rite emerges from and returns to a unity; the structure of the censing—a use of one symbol within the rite—emerges from and returns to a unity. The altar from which the hierarch emerges is not one point among many. It is the symbol of the intelligible stillness, from which Christ, and the rite itself proceed. This symbol only makes sense within the axial orientation of the church.

During the early portions of the rites, when the catechumens, penitents, and possessed are present, these groups and the laity apparently do not contemplate the altar itself, but the closed doors which lead to the altar area. The doors do not thwart the orientation toward the altar, since the rites of communion "have been well illustrated for the uninitiated on the doors of the inner shrine."[96] While there is some doubt as to what Dionysius means here, it seems that he expects that the doors to the altar area will have on them an icon of the last supper. The catechumens, who are "yet uninitiated in contemplation," may only contemplate the icon, and not the altar behind it. After the catechumens, penitents, and possessed have left the church, the doors are at some point opened, since the rite's culmination includes the elevation of symbolic objects above the altar for the contemplation of the laity.[97]

Though the laity form one group in the nave of the church, the consecrated orders have their own places, each one relative to the altar. The monks stand before the doors to the altar area, while the deacons, priests, and bishops stand within the altar area. During the consecration of the clergy, each is presented to the altar, and kneels before it. Dionysius explains the symbol: "the presentation at the divine altar and the kneeling outline to all the sacral initiates that they should submit their own life entirely to God, the principle of the rites, and offer their intellectual selves as worthy of the all-holy and sacred altar of the godhead, which sacrally consecrates deiform minds."[98] The kneeling of the clergy below the altar is a symbol of the submission of the mind to God. The altar here symbolizes God as consecrator. Dionysius elsewhere shifts from symbol to reality when discussing the kneeling of the deacons below the altar at their consecration, and treats the altar not as a symbol of God, but as the intelligible place of the mind's consecration. The order of deacons "places itself below the divine altar because purified minds are consecrated in it in an otherworldly manner."[99] The altar is the visible place of consecration for the order of deacons, who are sensuous creatures and so need a place for their consecration. The same altar, though, also serves as the intelligible place for the consecration of their minds and the minds of those they purify: the catechumens, penitents, and possessed. As both a sensuous and intelligible thing, the altar is a symbol of Christ, who is "our most divine altar."[100] In the church, we refer to the altar typically as the visible table standing behind the doors to the inner shrine. As Dionysius describes it, that table is simply the thing that marks out for our eyes the pivot between the sensuous symbolic things of the rite and the intelligible reality they embody. The hierarch may enter into thoroughly intelligible contemplation while standing at the altar because the altar is not simply the table. It is the table as visible ground for the intelligible context of the rite, and so it is the place in which the divine can appear, indirectly, through the symbols raised above it.

The altar as focal point serves also to ground the ranks of the ecclesiastical hierarchy. In his letter to Demophilus, Dionysius says that, during the rites, the monastic rank is stationed at the doors of the inner shrine. He seems to recognize that Demophilus will look for some role that they play in front of the doors, perhaps the role of guards, for he immediately clarifies that this is not the case: "they remain, not like sentries, but in order that each one be in his proper order and that they continue to be aware of being closer to the ordinary people than to the priestly ones."[101] The monks are stationed at the doors of the inner sanctuary so that there may be a ranking, and the altar allows this by constituting a center around which the ranks may form by their relative distance from it. The same ranking occurs in the funerary rites, when, "if the deceased belonged to one of the clerical orders, then he is placed at the foot of the divine altar," but if "the deceased was one of the holy monks or of the sacred people the hierarch puts him in front of the sanctuary at the holy entrance reserved for the clergy."[102] Dionysius later explains that these locations are symbolic of the correspondence between death and life. The rank a person holds in this life is preserved in the next. The altar again serves as the point of orientation for such an order, establishing a physical center to the sacred place, in relation to which the dead physical body may be placed.

God and the Intelligible Place

The altar is the "more divine place" within the larger sacred place of the church building. When the hierarch stands in front of it, and is "led up by the thearchic spirit to the holy principles of the rites in blessed and intelligible visions,"[103] does he stand in an intelligible rather than a sensuous place, or does Dionysius leave behind the language of place altogether when speaking of the intelligible? The lack of a sustained discussion of place in the Dionysian corpus impedes our work here, but Dionysius does explicitly identify intelligible places here and there throughout his work.

God as Place

Dionysius undertakes a thematic discussion of place only once in his entire corpus. The discussion runs only a few lines, and it mentions place explicitly only once. For all its brevity, it does articulate the relation of place to the divine names. There is no divine name of "place," though specific places are listed among the symbolic names.[104] Dionysius mentions place only in the course of discussing the paired names of "greatness" and "smallness," suggesting that place here will be defined as a kind of quantity or extension. He tells us first: "God is hymned in the scriptures as both great in size, and on a light breeze, which reveals the divine smallness."[105] The terms "greatness" and "smallness" are relative terms denoting quantity, and so seem to describe God as a quantity, but when applied to God, they refer to an unlimited, rather than a limited quantity. Otherwise God's quantity would be measured by something larger, and so he would not be prior to all things. As unlimited, God cannot be a passive extension, like the empty void described by John Philoponus. The greatness of God is active, and "gives itself to all great things. It is more than poured and more than extended outside of all greatness." The unlimited extension of God exceeds all finite greatness, and so becomes the limit of all things. The greatness of God, Dionysius says, "surrounds every place." God's greatness does not have place as the limit to his extension. He is himself place, as "that which surrounds" (περιέχον). By its unlimited extension, the greatness of God "crosses through all the unlimited in accordance with its more than fullness, its great works, and its fontal gifts." Dionysius provides here a wealth of technical terminology describing the emanation of relative things from the divine absolute, for which he has been variously praised or criticized, depending on whether the critic thinks emanation is compatible with Christianity.[106] For our purposes, it is enough to note that extension in our experience can only come about because of the unlimited extension of God.[107]

Once extension has become unlimited, it no longer needs a place to limit it, and so we find Dionysius occasionally telling us, as he does in the *Mystical Theology*, that God "is not in place."[108] In *On the Divine Names*, Dionysius says that the Trinity is not "in place, so that it could be absent from some place or cross from one place to another."[109] As with the Plotinian third kind of place, the fact that the Trinity is not in a visible place does not mean that it is absent from the cosmos. Visible places prevent the beings in them from being present in any other places. Absence from this kind of place means that "the Trinity is present to all things." This immanence of the

Trinity in all things must always be taken in tandem with its transcendence of all things. As Dionysius continues, "to say that it is in all beings comes up short, since the unlimited encompasses all things and is above all things." When we say that the unlimited is in all things, we must also say that it is not in all things, because it is above them. Otherwise we express only half the truth.

The Intelligible Place

The Hebrew scriptures describe several ascents of Moses up Mount Sinai to meet with God. Each time is slightly different, but once the scriptures mention that after Moses, Aaron, Nadab, and Abioud went up the mountain along with seventy elders of Israel, "they saw the place where the God of Israel stood."[110] Dionysius, like many of his exegetical predecessors, relies on the ascent of Moses as a useful allegory for the different forms of interaction between the human and the divine. A crucial point of the ascent involves the reaching of the summit of the mountain, which Dionysius describes in the *Mystical Theology* as the "place where God stood." To understand the nature of this place, we have first to separate it from the lower slopes of the mountain.

The various elevations of the mountain roughly correspond to the various levels of the ecclesiastical hierarchy.[111] Just as the catechumens, penitents, and the possessed may not take part in the rites of the church, since they are still awaiting the purification of baptism, so the impure are kept off Mount Sinai altogether. Moses is "himself commanded first to be purified, and again to be differentiated from those who are not so."[112] After his purification, Moses participates in a display of divinity which is both sensuous and multiple. He "hears the trumpets of many voices and sees many lights shining with pure rays pouring out in all directions." As in the church, so here, the sensuous participation in the divine is common to all of the laity. The nave of the church blends with the slopes of the mountain, as Moses and "the masses" together experience the sensuous manifestation of God. When Moses wishes to ascend to the peak, however, he must "be differentiated from the masses, and arrive with his chosen priests at the summit of the divine ascents." While the laity may stand on the slopes of the mountain, its summit is restricted to the clergy, like the altar area of the church. Here no sensuous display appears, but neither does God himself. Dionysius says: "he does not come to be with God himself. He does not contemplate him—for God cannot be contemplated—but the place where he stood." We may be tempted to think of this as a sensible place, perhaps the altar, where the bread and wine stand. But Moses has left all sensuous imagery behind, and the explanation given by Dionysius refers only to an intelligible structure: "I think that this signifies that the divine summits of visible and intelligible things are certain underlying structures (λόγοι) of things subordinate to the one who surpasses all things." These underlying structures are not God himself, since Dionysius has just denied that Moses contemplates God here. They are also not mere concepts, since Dionysius denies that concepts have any existence outside beings. The underlying structures of things are themselves beings, but the exemplary instances of beings.[113] For instance, the underlying structure of "life" is the most exemplary life. The highest of all lives, as Dionysius explains to us repeatedly, are the upper angelic orders. And indeed, in *On the Heavenly Hierarchy*, Dionysius refers to the angels as "the divine place of the godhead's rest."[114] It is the angelic minds, then, that give us the presence of God himself in the *Mystical Theology*: "through them, his presence above all conception is shown, walking on the intelligible summits of his most holy places." On this one occasion in his entire corpus, Dionysius uses the phrase "sacred place"—here "holy places"—to refer to something other than the church building. This is not to say that the contemplation of the "most holy places" occurs outside the church building. If we follow the liturgical imagery Dionysius has used in describing Moses' ascent, we can only conclude that the contemplation of the place where God stood is none other than the hierarch's contemplation of the intelligible structures while standing at the divine altar.

The hierarch, then, stands in two places at once. As the head of the ecclesiastical hierarchy, he convenes the members of the hierarchy within the church building, and he enters that place for their sake. By presiding over their gathering, he becomes the means by which they become united to each other, through the pervasion of the divine through them, and to the intelligible, through their orientation toward the altar and the symbolic things revealed there. The hierarch's procession into this secondary place through his many entrances with the censer, the gospel, and the bread and wine, goes hand in hand with his return, for his own sake, to the primary place: the intelligible structures that reveal the presence of the god who is beyond them. The hierarch's procession into the church leads the laity to contemplation, while his return to the altar restores him to his own contemplation.

Oppositions of Image and Symbol in Dionysius

Our discussion of place has so far left out the famous Dionysian ecstasy. By stopping our discussion of Moses' ascent with his contemplation of the divine place, have we neglected his important ascent beyond the summit of the mountain into the darkness above it? More importantly, does this location of darkness at the end of the ascent suggest that the sacred place is not so important to Dionysius? At the very least, some treatment of Dionysian ecstasy is in order here. It does not, in fact, stand far outside the place-oriented rites we have been describing.

Iamblichus has provided us with several examples of how ecstasy functions within his own place-oriented rites of Hellenic Neoplatonism. When used in the broad sense, ecstasy simply indicates the state of one reality when it performs the activities of a higher level of reality. Soul must efface itself, or become ecstatic, if it is to perform the activity of the intellect. When Iamblichus discusses ecstasy with Porphyry in *On the Mysteries*, he follows Porphyry in describing the more specific form of ecstasy found in rites of divination. Porphyry has seen forms of divination where "a self-effacement of reasoning" or "the mania that accompanies diseases" seems to be the cause of the divinatory power.[115] In short, the prophet seems to go mad in order to reveal the future. Iamblichus cautions Porphyry that the prophet experiences an ecstasy that is "above nature" (ὑπὲρ τὴν φύσιν), while the ecstasy of mania is "against nature" (παρὰ φύσιν). The prophet's self-effacement is not a mere privation of his self, but a suppression of his self so as to manifest a higher reality. While retaining the form of a human being, he acts symbolically, doing things that are either meaningless or impossible to his natural form. Iamblichus here combines his own theory of the symbol with the Plotinian language of self-effacement and stillness. The activity of the prophet "stills (ἡσυχάζουσα) its own life and understanding, and hands down the use of it to another."[116] The difference between Iamblichan and Plotinian stillness is simply that Iamblichan stillness charges with significance the rite that occurs around it, resulting in the announcement of an oracle to the participants in the rites. It thus reveals itself within a place, while Plotinian stillness affects only the mind of the person concerned, even if it ecstatically and salvifically encounters the One.

Dionysius never uses the term "ecstasy" in the specific divinatory sense of Porphyry and Iamblichus. He shows no interest at all in Christian oracles, though they certainly existed in parts of the world at his time.[117] He does, however, show an interest in the prophecy of the Hebrew scriptures, not as a source of foreknowledge but as a source of divine names. After discussing the origin of certain names in the unchanging providence of God, Dionysius goes on to say that some names may be drawn "from certain divine apparitions which illuminated the initiates and prophets in the sacred temples or elsewhere."[118] Dionysius never describes these apparitions as themselves ecstatic, though they may well have been the result of an ecstasy. He prefers to use the term "ecstasy" in the more Plotinian sense: the ecstasy of the pure mind, which stands in silence in the darkness of God himself after its thinking through of all things. Dionysius privileges the term for his hierarchs, Paul and Hierotheus, who, like Moses on the mountain, leave the laity behind when they contemplate without symbols the intelligible realities and stand in the darkness of God himself. The ecstasy of these hierarchs does not result in visionary writings, but it does seem to ground the ecclesiastical hierarchy. Dionysius says in the *Mystical Theology* that union with the divine darkness is beyond intellect, and that it is accomplished through an ecstasy.[119] The *Mystical Theology* describes only the ascent of the hierarch toward this union, and does not say whether it occurs in time, or whether its effect extends further than the union itself. But Dionysius describes a union beyond intellect elsewhere in his writings, and in these descriptions he does suggest that the union bears a fruit beyond itself. We have already seen how, in *On the Divine Names*, Dionysius describes what Hierotheus put into his book, the *Elements of Theology*. In addition to his own intellectual labor, and what he received from the theologians of the past, Dionysius says that Hierotheus also included "whatever he was initiated in by a certain more divine inspiration," an inspiration which involved Hierotheus' being "perfected by a mystical union."[120] Hierotheus was united with "divine things," and consequently was able to put these divine things into his written work. The form that these divine things take in his work is visible (since they are written words), their content is intelligible, but the mode of their reception goes beyond the senses and the intellect to a form of union. Even the angels charged with handing down the intelligible content of the rites to the hierarchs must first receive it themselves through a union that goes beyond intellect. If their angelic intellection did not require a union with the divine, then their thoughts could no longer be salvific of either themselves or the human beings to whom they give their thoughts in the form of the rites. The hierarch, who is himself at home in the intelligible, seems to acknowledge that he is not the equal of the rites he performs when he acts "in a manner handed down by God," and even "explains, sacredly crying out to God, his performance of a sacred work which is above him."[121] Dionysius does not tell us the character of the union that grounds this sacred work. He uses the language of sensation at some times, the language of thought at others, but in general he does not let us know for certain whether this union is an experience beyond intellect or simply our acknowledgement at the level of intellect that what we know is beyond us. In either case, the rites are ecstatically grounded.

Dionysius suggests in one passage that the cosmos, too, is ecstatically grounded. He tells Timothy that "the author of all things himself, by his providence for all things, comes to be outside of himself through the transcendence of his erotic love, by a beautiful and good eros for all things."[122] The author of all things comes to be outside himself by falling in love with the world that he himself creates, apparently by this same self-effacement. If God creates all things by coming to be outside himself, it seems possible that the cosmos may be nothing but this God-outside-himself. Dionysius, however, goes on to say that God "is as though charmed by goodness, love, and eros, and descends into all things from being removed from all things and beyond all things, by his self-effacing super-substantial power, which cannot be stretched beyond him."[123] With this last clause Dionysius confines himself to the world of early medieval Neoplatonism, where causes do enter into their effects, but always carefully preserving the difference between the cause and the effect. In other words, the substance of a creature can never be strictly identified with the substance of the creator. As Dionysius often says, substances (*οὐσίαι*) are knowable, and the author of all things is not.[124] He is, then, above substance (*ὑπερούσιος*). We cannot consider a creature as having the unknowable God as its substance, hiding behind its visible and comprehensible attributes. This understanding of knowing is typical of the early medieval Neoplatonists, and reveals the inheritance of Aristotle, who himself says that "the attributes contribute a great deal to knowing what a thing is."[125] Visible properties reveal the substance of the thing whose properties they are. It awaits later readers of Dionysius, such as Eriugena in the ninth century and Meister Eckhart in the fourteenth, to see the substance of the thing as unknowable and radically removed from its properties. They begin to see the substance of the cosmos as God-outside-himself, and so itself a sacred place filled with sacred things.

The ecstasy of the human hierarch results in human actions that are meaningless when considered according to the nature of a human being. The things used in the rites manifest a symbolic character, which is distinct from their form, though not opposed to it. Dionysius restricts actions that are impossible by nature, when the symbol opposes the natural form, to God. He gives two examples of such actions: the resurrection of the dead and the miracles of Christ. God seems not to transcend nature but to go against nature when he promises to imbue our human souls and bodies with an immortal life. Dionysius acknowledges that "this event seems to antiquity to go against nature," but he responds: "to me and you and to the truth it seems divine and above nature."[126] Iamblichus used the phrase "above nature" to describe actions performed by animals, and birds in particular, which directly contradict their natural instincts—when a bird dives into the ground, for instance. When an animal does something that is meaningless by nature, we know that something divine is at work in it. The idea of our animal selves possessing an immortal life seems impossible to our nature as animals. As animals, we are inseparable from our bodies, and bodies die. The immortal life of an animal is then not natural, but "above nature." Though Dionysius uses the phrase "above nature" to describe immortality, he prefers to restrict it to our knowledge of nature, and not to use it as a description of nature itself. We may call immortal life "above nature," but from the perspective of divine life even this is not above nature, since the divine life "is the nature of all living things, and especially of the more divine among them." If we may draw a conclusion from this slender example, it is that Dionysius is sensitive to the difficulties posed by the late Neoplatonic bifurcation of the world into symbols and images. The divine is cause of both symbol and natural form, and so their opposition to each other seems unaccountable. Dionysius suggests here that the opposition is only apparent from our perspective. From the divine perspective, symbol and image are equally natural.

In our present life, we see things performing actions impossible by nature only in the case of the miraculous. Dionysius does not engage in hagiography; we hear of no miracles performed by the saints. He discusses only the miracles of Christ. In a letter to the monk Gaius, Dionysius says of Christ: "a virgin supernaturally gave birth to him, and unstable water did not give way, but held up the weight of his material and earthly feet by a supernatural power."[127] The term "supernatural" here indicates that natural things perform feats that are impossible by nature in the presence of Christ. Rather than exhibiting their natural forms, they become symbols of Christ's supersubstantiality. Dionysius makes the symbolic character of these actions clear in the next lines, where he says: "every affirmation regarding Jesus' love for humanity has the power of a transcendent denial." Here he means by "affirmation" the positive statement of a miracle, such as: "he came to them, walking on the sea."[128] The force of the statement lies in the impossibility of its content. The impossibility gives it the force of a denial, and this denial is its meaning. What seems to be a man walking on water is not: it is someone divine. Dionysius here incorporates the miracles of Christ into a larger symbolic context, which includes the rites of the hierarchy. The contemplation of bread and wine, too, is meaningless when their nature is considered, but becomes an interaction with the divine in the mythic context of the rite.

CHAPTER 4

THE DIONYSIAN TRADITION

The Dionysian writings appeared first within the Christological controversies of the sixth century. It was long thought that they were first cited by the Monophysite party, whose doctrines were condemned as heretical by the Council of Chalcedon in 451, and so the Dionysius writings were frequently assumed to have been written by a Christian heretic. Paul Rorem and John Lamoreaux have recently shown that all parties in these disputes, both Chalcedonian and Monophysite, took up the Dionysian corpus with vigor at the same time, and so we do not need to assume a heretical milieu for their author.[1] The corpus soon attracted a great deal of interest even outside the context of Christology, and received its first commentary within fifty years of its composition. This commentary, now known as the Dionysian scholia, has long been associated with the name of Maximus the Confessor (580–662), though it now appears that he was responsible for few, if any, of the scholia.

Whether or not he wrote any scholia on Dionysius, Maximus the Confessor read the corpus carefully and enthusiastically. Inspired by the *Ecclesiastical Hierarchy*, he undertook his own explanation of the church and its rites in the *Mystagogy*, a work that explains the Christian rites in terms of the church where they are performed. While Dionysius may have inspired Maximus to reflect on the nature of the sacred place, the Dionysian hierarchical ordering of creation, and consequent rationale for why certain humans need the sacred place, had less influence on Maximus. After Maximus left his early career at the Byzantine imperial court, and took up residence in nearby monasteries—first Chrysopolis, then Cyzicus—he developed an understanding of the human being as "microcosm" or "little world," containing in itself every level of the hierarchy that Dionysius parcels out to different species. Gregory of Nyssa had famously criticized the term "microcosm" as unworthy of human nature.[2] To him, the definition of the human as a "little world" implied that the changeable, corruptible character of the visible world was now characteristic of human nature as well. Maximus is able to accept the term "microcosm" because for him it does not mean the inclusion of all levels of reality in the human, but that human nature stands prior to all the levels of reality, from the lowest to the highest. Neither the place of all things nor something in place, it binds together the places (that surround) and things in place (that are surrounded).

We do not find Maximus' mature formulation of the microcosm theory until the writing of his *Ambigua ad Iohannem* (628–30), or "Problems for John," written apparently as a record of conversations he had with his superior in the monastery at Cyzicus.[3] One of the *ambigua* discussed by Maximus and John is a short passage from the work of Gregory of Nazianzus: "natures are made new, and God becomes man."[4] They wonder what Gregory could mean when he says that Christ's incarnation has "made nature new." Before Maximus can explain Gregory's meaning, he must describe how nature was originally made. This entails a description of nature's five divisions, all of which turn out to be held together by human nature. The divisions are as follows: (1) nature may be divided into uncreated and created; (2) created nature may be divided into intelligible and sensible; (3) sensible nature may be divided into heaven and earth; and (4) earth may be divided into paradise and the inhabited world. Human nature stands prior to each of these divisions as "the workshop that holds all things together."[5] The fifth division, by which human nature is itself divided into male and female, does not have the same structure as the first four. A result of sin, it occurs within the human, dividing human beings from each other. The first four divisions of nature follow the model of neither sensible nor conceptual division. Sensible division is always of a whole into its parts, for example, the whole loaf of bread into individual slices. The division of concepts may, on the contrary, be a division of a genus into its species. We may, for instance, divide the genus "animal" into the species of "human," "deer," "bird," and so on. The species "bird" is not a part of the whole "animal" in the way that a slice of bread is part of the loaf. A slice of bread is not a loaf, but a bird *is* an animal. That is, a bird is wholly an animal, and not merely a part of the whole.

The fourth division of nature seems at first to follow the sensible model of division. The earth seems to be divided into paradise and the inhabited world just as a loaf of bread is divided into slices. Yet, as Maximus will later make clear, paradise is not a different part of the earth from the inhabited world. Paradise is the only part of the world we ought to inhabit. If we inhabit somewhere other than paradise, we lack the "completion" (*συμπλήρωσις*) of our habitation, which comes about only when human nature overcomes the division between paradise and the inhabited world.[6]

The division here is not of a whole into its parts, but of an activity into its actual and potential state. Our inhabited world is potentially paradise; it is most fully actual when it becomes paradise. Only then do we see habitation in its most perfect sense.

The third division of nature into heaven and earth also seems at first to divide a whole (the sensible cosmos) into its parts (heaven, meaning the celestial bodies and the region through which they move, and the earth), but for Maximus as for his predecessors, the heaven and the earth are not related like two slices of bread. The motion of the heavenly bodies is ultimately responsible for the production of all motion on earth. All motion on earth, in turn, strives to resemble the motion of the heavenly bodies. That is, the heavens are actually what the earth is potentially: in perfect motion. Here again, we see that the division of nature is a division between actuality and potency. Maximus occasionally refers to heaven and earth respectively as "what surrounds" (*περιεχῶν*) and "what is surrounded" (*περιεχομένον*),[7] using the language of place and what is in place to characterize the two poles of this division of nature.

The first two divisions of nature continue, at least implicitly, to employ the pattern of actuality and potency, and to think of the actuality as the place of the potency. The second division of nature divides the created world into a sensible and intelligible creation. Maximus does not here say that the sensible is the intelligible in potency, while the intelligible is the sensible in act, but he does say so elsewhere. In the *Mystagogy* (628–630), for example, he says that the church's nave and sanctuary are images of the sensible and intelligible substances respectively, because "the nave is the sanctuary in potency. . .and the sanctuary is the nave in activity."[8] Just as the activities in the nave depend on and are completed by the activity at the altar, so the sensible depends on and is completed by the intelligible. The intelligible already is what the sensible is always becoming. The first division of nature takes the intelligible substance, in which the sensible finds its actuality, as its first pole. As the epitome of creation, this intelligible substance comprises beings that are "quite worthy and superior to being related by genus to the one who is quite without relation, and are naturally freed from all things."[9] These intelligible beings are not related by genus to the uncreated God, who "is quite without relation." They instead overcome their division from God by becoming without relation themselves. In his *Chapters on Theology*, written around the same time as the *Ambigua*, Maximus describes this abandonment of relation as the taking of God as one's place. One who receives God "will be above all ages, times, and places, having God himself as his place."[10] Place, in the ordinary sense, belongs with age and time "among the relative terms," but "God is not among the relative terms." Instead of place in the ordinary sense, the saint in whom the four subsequent divisions have been overcome is united to God as his place. These references in Maximus' writings contemporary with the *Ambigua* reveal that the concepts of place, potency, and actuality remain implicitly present in the first two divisions of nature, though Maximus does not refer to them explicitly, perhaps because they become less precise and more metaphorical as we advance farther beyond the sensible.

Were we still relying on a Plotinian understanding of human nature, we would conclude from these divisions that we as human beings should strive to identify ourselves with the surrounding element in each. We would strive to be paradise, then heaven, then intelligible, then God. As we have seen, Plotinus understands our higher nature to *be* that surrounding element, so our goal would simply be to identify ourselves with our true self. Gregory of Nyssa denied that we should identify ourselves with the "microcosm" because he accepted this Plotinian account: that we are not essentially sensible or earthly. Maximus, on the other hand, introduces a new understanding of human nature to the Neoplatonic tradition. He identifies the human neither as what is in place, as we find within the Iamblichan and Dionysian tradition, nor as (the intelligible) place itself, as we find in Plotinus, but as prior to both place and what it contains. Without human nature, neither place nor what it contains could exist, since the human is nothing other than the bond that holds the two together. Human nature "is the workshop that holds all things together, and naturally mediates, through itself, between all the poles of each division."[11] By nature, each of us already has "the property of relating to all the poles in our own parts." Our own parts bear a relation to each pole of the divisions of nature, again with the exception of the fifth, unexpected, division into male and female. The mediating of the divisions, however, requires an act on our part, since by nature we do not have the act, but only "a natural power for union with the mean point of all the poles." By actively uniting the poles, the human being is "established clearly as the great mystery of the divine aim." Without human nature, the cosmos is radically incomplete, not simply missing one of the links in its chain, as it would in any of the philosophical systems we have already surveyed.

Although human nature is prior to the divisions of nature, the human beings who are charged with overcoming its divisions find themselves immersed in one or another of its poles. The overcoming of the divisions in nature, then, begins with the overcoming of the divisions in themselves. A male overcomes the division into male and female "by a most passionless relationship to divine virtue," which eliminates the property of maleness in him.[12] As men and women, we desire each other sexually. If we cease to feel this passion, we cease to exhibit the property of this division. Like his predecessors in the monastic tradition, Maximus does not regard the elimination of passion as an easily achieved act of the soul. It requires ascetic practice. More than a single act of mediation is also required to overcome the other four divisions of nature. Since these divisions are not the result of sin, their poles are not to be destroyed altogether. Instead, the lower pole is to be assimilated to the higher pole. We overcome the division of paradise and the inhabited world "through a way of life proper to and fitting for the saints."[13] Instead of leaving the inhabited world or attempting to transform it directly, we change our way of life within that world, and so we experience (παθόντι) the world differently, "not divided by the difference of its parts, but gathered." Paradise is the proper place of the inhabited world, gathering it together so that it does not disperse into its various parts. Our ascetic transformation of ourselves puts the inhabited world in its place.

Maximus continues to use the language of place and ascetic transformation in his description of the division into earth and heaven. We overcome this division not by leaving the earth or transforming it, but by "a life of virtue in every way like the angels."[14] Our ascetic practice causes us to become "light in the spirit, and held down to the earth by no bodily weight." We do not cease to possess bodies, but we cease to identify ourselves with them by living a life not oriented toward the body. As a result, our experience changes. The sensible creation no longer "divides itself into places by any extensions at all." Instead, it acquires the nature of spirit—the material of the visible heaven—that is undivided. As it did with the inhabited world, our way of life puts the earth into its place. The earth is no longer divided into a plurality of places, but is made one.

The overcoming of the last two divisions of nature directs the sensible world to something beyond it, and leaves behind the language of place and ascetic practice. The pattern, however, remains the same: we must adopt the way of life or thought belonging to the greater pole. Because we, in every case, start off immersed in the lesser of the two poles—the inhabited, earthly, sensuous, created world—Maximus can still find a place for the visible rites of the church, even though he does not, like Iamblichus and Dionysius, involve our *nature* in these poles. The church may become the place of our transformation, established by Christ, who undertook in himself the first and exemplary overcoming of nature's divisions.

By the time Maximus came to write the *Ambigua*, he had reluctantly left Cyzicus for what he hoped was a short period of exile in North Africa. As we will see, Maximus' understanding of place comes into play in letters he writes from this exile. His microcosm theory, as set out in the *Ambigua*, however, constitutes the foundation for his thinking about place, as well as that of his major exponent in the Latin West: the ninth-century Irish philosopher Eriugena. Eriugena very likely did not read either the letters or the *Mystagogy* of Maximus, but he carefully read and translated the *Ambigua*, and incorporated Maximus' microcosm theory in his own work.

Maximus the Confessor

Between 630 and 633, Maximus wrote a work entitled *Questions for Thalassius*, apparently written to a new interlocutor he met while in exile in Africa. Of the sixty-five questions Maximus set himself to answer in this work, one reads: "if God does not dwell in a temple made by hands, how did he dwell in the temple of the Jews?" Maximus gives what at first seems to be a thoroughly Dionysian answer to the question. He says: "God dwells in the temple of the Jews figurally, but not truly, circumscribing his ineffable will by this dwelling in the temple. It is an education which thoroughly initiates his charges."[15] God does not enter the Jewish temple in himself, but only through the mediation of a figure. By employing a figure, God is able to initiate the Jews, who require that he dwell in a visible place. Maximus goes on to say: "only the pure intellect is fit to be the dwelling of God. On account of this, he consented to inhabit the figural temple, wishing to draw up from matter the debased mind of the Jews through the baser symbols to the much more invisible figures."[16] Two less Dionysian ways of speaking begin to become apparent here. The first is the telescoping of the human and angelic intellect. Dionysius would agree that "only the pure intellect is fit to be the dwelling of God," but his pure intellect belongs only to the highest ranks of angels, who are fit to be "the divine place of the godhead's rest."[17] The human intellect is by its very nature not as fit to be exercised directly. It needs visible figures, then, not because it is "debased," but because it is human. Maximus' second departure from Dionysius here is his treatment of the temple building itself as a figure. Dionysius treats the church building as the place where things become symbolic, and he calls it "sacred," but he never treats the building itself as a symbol. Maximus here continues a line of thought he develops in his slightly earlier *Mystagogy*, in which he describes the church as an image of God in its overcoming of the various divisions of nature. In the *Mystagogy*, Maximus does not describe the church as a concession to our immersion in the sensible, but by identifying its activity with an activity already occurring in the cosmos as a whole, he suggests that we overcome the divisions of nature in the church only because we are unable to overcome them in the cosmos. The distinction between place and sacred place becomes strained here, in that the sacred place no longer has a character radically different from that of place.

Place and the Divisions of Nature

In a portion of his *Ambigua* devoted to the contemplation of nature, Maximus briefly takes up the nature of place. He aims to demonstrate that "everything whatever save God is entirely in place."[18] Maximus does not emphasize the distinction between the Aristotelian terms "where" (πoῦ) and "place" (τόπος), but he does suggest that "where" only becomes relevant when something is already in place. When we can answer the question of "where is it?" we know that something is in place. Maximus does not demonstrate that each kind of thing in particular has a place, but leaps straight to the question of whether there is a place for the "totality of things" (τὸ πᾶν). This is not the question of whether the cosmos has a place. The totality described by Maximus here is not a visible totality, like the cosmos, but the totality of everything visible *and invisible.* Aristotelians, and certain late Neoplatonists like Damascius and Simplicius, allow places only for bodies. Maximus provides a more Plotinian or Iamblichan treatment of place, since they allow for places of both visible and intelligible things, but Maximus continues to use Aristotelian language. The place of the totality must be the surrounding limit of the totality. For Maximus, there must be something beyond the totality that can serve as its place because "the totality itself is not beyond the totality of the totality. It would be somehow both irrational and impossible to say that the totality itself is beyond the totality of itself."[19] The totality cannot be self-limiting, but whatever turns out to be its place will have to be, since "the limit itself is outside itself, having its circumscription from itself and in itself, after the unlimited power of the total cause, which circumscribes all things."[20] Maximus seems to nominate God as the place of the totality, since God's unlimited power allows him to be self-circumscribing. Having concluded that there is a place of the totality, Maximus does not go into the specifics of this place and its identity with God. He simply sums up his definition in Aristotelian language. The self-circumscribing limit "is the place of the totality, just as some define place, saying that place is the periphery outside the totality, or the position outside the totality, or the limit which surrounds what is surrounded in as much as it is surrounded."

When Maximus takes up the divisions of nature in the *Ambigua*, he articulates a more complex relation between God and place. Maximus has said that human nature is appointed to overcome the divisions of nature. A division is overcome when the beings within that division become complete, actual, and perfect. The division of the earth into paradise and the inhabited world is overcome when the inhabited world becomes like paradise, the perfection of the earth. The division of the sensible into heaven and earth is overcome when the earth becomes like heaven, the perfection of the sensible, and so on. We may, then, say in an imprecise and restricted sense that unity in nature comes about when species (heaven and earth) are united in their genus (heaven as sensible creation), or when parts (heaven and earth) are united in their whole (heaven as sensible creation). The genus and the whole are the place of the species and the part, just as heaven is the place of the earth. The imprecision in such language quickly becomes apparent. The Neoplatonists, following Aristotle, are willing to say that species are "in" their whole, but they typically distinguish these kinds of "being in" from being in place. Maximus will use the language of place to describe both whole and genus, saying that the whole is what gathers the parts into itself, but he will also describe place as what gathers the parts into the whole while remaining distinct from both.

God plays the role of both kinds of place. On the one hand, he is "the undefined and unlimited definition and limit of every. . .nature."[21] God is the higher pole of the first division of nature, the division into created and uncreated. Once the lower poles of the earlier divisions have achieved their limit in the higher poles, it now seems that creation itself must achieve its limit by becoming God. On the other hand, in the person of Christ, God holds together the poles of each division, rather than constituting a pole himself. Maximus explains that all things have "internal structures" (λόγοι), both the things that are genera and the things that are parts of genera. A particular rosebush has internal structures that allow me to recognize and understand it: it is of a certain height, has a certain shape, and so on. These internal structures (λόγοι) allow me to come up with the definition (λόγος) of the rosebush, and name it using the word (λόγος) "rosebush." There are more general structures that the rosebush participates in, though they are not specific to the rosebush. Maximus is not always consistent about the number and character of these more general structures, but his most thorough treatment of them identifies them as being, identity, difference, position, and mixture, a variation on the five traditional Platonic genera.[22] Christ, who is "the wisdom and prudence of our God and Father," holds together the particular structures with each other, the generic structures with each other, and he binds the particular to the generic so that the two become one.[23] "The structures of the more generic and universal beings are held together by wisdom," and "the structures of the parts. . .are surrounded by prudence."[24] Christ here plays the role of place, holding together and surrounding the various structures of creation. He overcomes their division from each other when the structures of particulars "abandon the symbolic variety in their material things, are unified by wisdom, and receive a shared nature in their identity with the more general structures." As wisdom, Christ brings the more divided structures (the lower pole) into identity with the more unified structures (the higher pole). Christ himself is not one of these structures, not even the highest of them, even though the Gospel of John describes him as "the word" (λόγος). Instead, he binds the created structures together, while as God, he remains outside them. There is a tension, then, in the language of place used by Maximus to describe God. God overcomes the divisions of nature because he is the place of nature, while he remains separate from the things he gathers into himself as place. Yet God is also the highest pole of the divisions of nature, and so he overcomes these divisions only when all things are united in him as their own actuality. This tension is not to be overcome in Maximus, and is his own formulation of a tension we have seen since Plotinus in the Neoplatonic highest place, which must both be all things and be beyond all things.

Maximus' *Mystagogy*, written around the same time as the *Ambigua*, contains the same tension in the presentation of God as place, though here the tension extends itself to the visible sacred place: the church. The *Mystagogy* continues to ascribe to God the characteristics of place. Maximus reminds the reader that God not only brings all things into being, both intelligible and sensible things, but also "holds them together, gathers them, and circumscribes them, and he providentially binds them to each other and to himself." God is able to bind all things to each other through what Maximus in the *Ambigua* calls the "one thoroughly undivided structure" possessed by every created thing.[25] This is precisely the structure of "production from non-being." Every created thing is united to every other created thing by the fact of being created, or having a cause. Because of this common character, Maximus can say in the *Mystagogy* that God "makes things distant from each other in nature to acknowledge each other by his single power of relating them to him as cause." Though different from each other, the things in the universe are the same in that they are related to God as cause. This connection with each other leads them "to an unconfused identity of motion and subsistence."

The nature of the "unconfused union" is a central concern for Maximus throughout his work.[26] The term pops up now and then in Neoplatonic literature, but comes to prominence in the Fourth Ecumenical Council, the Council of Chalcedon, which uses the term to describe how the divine and human nature come together in Christ. The two natures remain "unconfused," even though they occur within a single person. Maximus amplifies the scope of this term so that it applies not just to the historical person of Christ, but to the interaction of created things generally with God. Christ's human nature and the totality of things relate to the divine nature in the same way: unconfusedly. Maximus goes on in the passage from the *Mystagogy* to describe how all things come to an unconfused unity in

God. The single relation established by the first principle "makes vain and conceals the proper relations contemplated according to the nature of each being, not by corrupting and destroying them and making them not be, but by conquering and more than revealing them, as the whole does the parts." In this last clause, we find a clue as to the meaning of "unconfused unity" in Maximus. We do not have to look to Christ for an example of it; we see such examples every day. The parts of every visible whole are also visible, but we do not attend to them when considering the whole. When I look at my watch to see what time it is, I consider the watch as a whole, and never pause for a separate consideration of the different hands, the face, the band, and the rest of its parts—though these parts do not cease to exist even when I only consider the whole. I give the watch what Maximus and the Neoplatonists generally call a "comprehensive glance" (ἀθρόα ἐπιβολή).[27] The same goes for the totality of things when they are contemplated in God. They do not cease to exist, and they maintain all their own characteristics, but we see those characteristics only in our vision of God himself. God is the whole, which binds the parts together into one.

The language of whole and part seems to suggest that God is no more than creation in its integrity, a conclusion Maximus wishes to avoid. After he says that the first principle "makes vain and conceals the proper relations contemplated according to the nature of each being,. . .as the whole does the parts," he goes on to say "or as the cause is revelative of wholeness itself, by which the wholeness itself and the parts of the wholeness are revealed and come into being. The cause shines beyond them." God is both the whole, which gathers all creatures into itself as parts, and the transcendent cause of both whole and part. The division within the divine place helps to explain why Maximus can say meaningfully that the church is both "the figure and image of God" and "the figure and image of the world."[28] If God is the place of the cosmos, it seems at first contradictory that the church could be the image of both God and cosmos, since it would then act both as what surrounds and what is surrounded. But God is place in two ways. He is the whole that gathers the parts into himself, and he is the force that gathers the parts into the whole without identifying himself with either one. In this latter case, the whole is not God himself, but the cosmos in its integrity, acting now as the place of all its parts. It is in this sense that the church can be the image of both God and cosmos. Like the cosmos, it gathers together its parts into itself as a whole. But, like God, it also binds together its parts into a whole that it does not itself constitute. In both these capacities, the church allows human beings to participate, at a local level, in the overcoming of the divisions of nature that they can no longer accomplish in the cosmos as a whole.

The Sacred Place and the Divisions of Nature

William Ramsey, whose turn-of-the-twentieth-century explorations of Turkey yielded some of the first evidence of continuity between Hellenic and Christian rites, denies that Christianity could have remained Christian if it had adopted the Hellenic practice of making rites specific to the location where they were performed. "Local variety," he says, "is inevitably hostile to the Christian spirit, because Christianity is unity, and its essence lies in the common brotherly feeling of the scattered parts of a great single whole."[29] Maximus brings a similar spirit to his theory of the sacred place. Where Dionysius treats the walls of the church as the boundary within which things can become symbolic, and the content of the symbolism is invariably intelligible, Maximus treats the church itself as symbolic, and the content of its symbol includes both the sensible and intelligible components of all creation. Every church building is simply the localized symbol of the whole creation. Dionysius follows the Hellenic use of "symbol" to indicate a sensible thing with a special tie to intelligible gods. Maximus returns to the language and meaning of "figure," a term which primarily indicates a sensible thing that adumbrates another sensible thing. Typically used by Christians to describe events in the Hebrew scriptures as the foreshadowing of events in the life of Christ, "figure" in Maximus' *Mystagogy* allows parts of the visible church building to become the symbol of the whole visible creation.[30]

"The holy church of God," Maximus says, "does the same things for us as God, and in the same way, though as an image to its archetype."[31] It is not that the church replaces God, but that we humans, who are no longer able to share the activity of God in the cosmos as a whole, now share in it through the mediating activity of the church. The church allows the activity of God in the whole to manifest itself symbolically in a part of that whole. This is the ideal thing for a creature that has "misused its natural power by the division of what was united," as the *Ambigua* says, and now finds itself oriented toward the part rather than the whole.[32] Christ not only became incarnate to take over the mediation of the divisions of nature that human beings had abandoned, but he also established the church as a new arena of mediation for human beings. The "men, women, and children" who come into the church are "divided from each other by genus and species, peoples and languages, ways of life, ages, orientations, crafts, methods, customs, suitabilities, knowledge and presuppositions, accidents, characters, and dispositions."[33] The church, however, "gives one thing to all of them,. . .to be and to be called after Christ." The church bestows a unity on the people who enter it which is more than the adventitious unity given by their mere presence within the same walls. The people receive a "divine shape," by sharing in the activity of Christ. Maximus builds on the apostle Paul's metaphor of Christ and the church as a body and its members, saying that the people are "one body of many members, worthy of Christ himself who is our true head."[34] Acting as a whole, then, the church overcomes the first division of nature by reuniting creation, embodied by human beings, in God. In its parts, however, the church also acts to overcome the two subsequent divisions of nature into the intelligible and sensible creation, and into heaven and earth.

The parts of the church building come into explicit play for the first time in the second chapter of the *Mystagogy*, where Maximus explains the roles of unity and difference in the unconfused union he has described in the preceding chapter. The world and the church both constitute an unconfused union. They are capable of it, Maximus says, because they allow in themselves "both union and difference."[35] The church building has unity and difference in its very architecture. It is one building—one entity or *hypostasis*, Maximus later says—but there are different positions within that building. The positions are two: the sanctuary (*ἱερατεῖον*) and the nave (*ναός*). Following the ritual practice that occurs within the church building, only the priests may enter the sanctuary, while the whole body of the faithful may enter the nave. These positions do not then differ arbitrarily, as though they were merely different in shape, size or relation to the exterior walls. The difference within the church building is not merely architectural, but ritual, bound up with what happens within the different positions. Now if there were only difference between the two positions, the church would cease to be a single entity, but they are united by the building itself, which "frees the parts from the difference in their name by its elevation to unity, and shows them to be the same as one another." Maximus perhaps overstates his case here, as he shortly explains that the two are not exactly "the same as one another," but are related as potency to act. A child is an adult in potency; the adult is the actualization of the child. They are not simply two different entities, united by their common species. Their unity lies in the fact that they are the same entity—one *hypostasis*—separated in time by degree of actualization. The child is in the constant process of becoming the adult. Like child and adult, the nave and sanctuary differ by degree of actualization—"the nave is potentially the sanctuary, since the relation of the initiation to its limit sanctifies it"—but this difference is manifested in position, not time. Because they are not separated in time, the nave can never actually become the sanctuary, yet they remain the same entity. It is something like the relation between the soul and body of a child. The soul has the principle of the adult in it, but the body does not fully realize that principle. The soul, then, is the actualization of the body, because it possesses the principle of the body's actualization. In the same way, "the sanctuary is the actualization of the nave, since it possesses the principle of the initiation proper to it." But just as without the body there can be no entity, so without the nave there can be no church.

The relation between act and potency in the various parts of the church allows Maximus to identify the church as the place where the divisions of nature are overcome, since these are themselves divisions of act from potency. Maximus tells us that, although the "whole cosmos of beings" is divided into an intelligible cosmos and a sensible cosmos, it is "one cosmos, not divided by its parts."[36] The cosmos avoids becoming an aggregate of different parts because its parts are not merely different, but in one another. The "whole sensible world subsists in the whole intelligible world intellectually, when it is simplified by its structures." The structures of the sensible world are intelligible; they are what make the sensible world comprehensible. If it did not subsist in the intelligible world, the sensible world would fall apart, no longer having any structure. But the sensible world is not simply dependent on the intelligible. The sensible world makes the intelligible world visible. "The whole intelligible world appears in the whole sensible world when it is mystically figured by symbolic forms." Here as elsewhere, Maximus freely blurs the distinction between symbol and natural form. The sensible cosmos is itself a symbol of the intelligible cosmos. We are no longer able to make use of this cosmic symbol, and so we enter the church, where the Dionysian hierarchy still applies. The hierarch contemplates the intelligible structures of the sensible world in the sanctuary, while the laity see the appearance of the intelligible world in symbolic forms. Through their shared activity in the rite, the division between the sensible and intelligible is overcome.

The overcoming of the division between heaven and earth receives a laconic treatment in the *Mystagogy*'s third chapter. Maximus tells us that the church has "the divine sanctuary as heaven, and the good design of the nave as earth."[37] It is not clear whether the sanctuary and nave here share the activities of what they symbolize, as do the symbols of the first two chapters, or whether their symbolic content is foreign to them, having meaning only within the context of the rite, as is more typical of symbols later in the *Mystagogy*. Maximus says little about them, giving only the single clue that the nave has a "good design" (*εὐπρέπεια*) and the earth has an "order" (*διακόσμησις*) to it. The nave is a larger, more differentiated area, and reveals a greater concern for proportion between its parts than the sanctuary, which is smaller and less differentiated. To the degree that simplicity rather than proportion characterizes the sanctuary, it resembles the simplicity of the heavenly element of fire. In this regard, Maximus gives to the sanctuary the character that Dionysius gives to the altar.

Dionysius uses the term "inner shrine" (ἄδυτον) for Maximus' "sanctuary" (ἱερατεῖον), and he rarely speaks of it, preferring instead to focus on the altar itself. When the Dionysian hierarch leaves the altar, he moves from the unity of the intelligible to the multiplicity of the visible. Since Maximus makes the larger area of the sanctuary the symbol of the intelligible and heavenly, we may well wonder what he reserves for the altar. Maximus brings the altar into his discussion of the church building as an additional level of refinement in chapters four and five. We now have three positions to distinguish within the church: altar, sanctuary, and nave. The nave is the place where bodies take on the proper relation to each other (through "ethical philosophy," Maximus says). The sanctuary is the place where the soul engages in contemplation, the act Dionysius reserves for the altar, and the altar is the place where the mind engages in mystical theology. The phrase "mystical theology" comes from Dionysius, who says in his work of the same name that the "truly mystical darkness" is entered by Moses after he sees and hears the sensible manifestations of God on the slopes of Mount Sinai and contemplates the intelligible structures on its summit. We have seen how the Dionysian hierarch follows these same steps within the sacred place, though Dionysius does not reserve a special place for entry into the mystical darkness beyond the sensible and intelligible. The contemplation of the intelligible principles of the rites and the entry into the mystical darkness both take place at the altar. Maximus restricts the altar to the mystical, where the mind "invokes with a different speech and many-voiced silence the many-hymned silence of the hidden and unknowable great voice of the divinity in the inner shrines."[38] At the altar, the mind prays to what is beyond it, and so silence is its appropriate form of address, since words always grasp at least some part of what they signify. This silence is not simply the absence of speech, which happens in mundane places far from the altar. It is the transcendence of speech, and so Maximus calls it "many-voiced," meaning that it says more than the spoken word can. This activity at the altar continues in the work of Maximus the ecstatic grounding of the rites of the church described by Dionysius. The church "gathers all things to the mystery completed on the divine altar."[39] This is the church's overcoming of the first division of nature—when the creation made one within the church building is offered to the God who is beyond it, an offering that is properly accomplished only in the mind.

Were human beings not oriented toward the divided, they could accomplish in the cosmos as a whole the same tasks they now perform in the church building. Borrowing from the apostle Paul, Maximus says that the cosmos is "a church not made by hands," which is "wisely manifested through the one made by hands."[40] This styling of the cosmos as a church strains the boundary, so important to Dionysius and the later Hellenic Neoplatonists, between natural form and symbol. The forms of things in the sensible and intelligible worlds have themselves the character of being within a great cosmic church. We are simply unable to elevate these forms to their divine source. Like the Jews described by Maximus in the *Questions to Thalassius*, our "debased intellects" are unable to unify the whole of the cosmos and offer it to God. We need the visible church, made by hands, which offers us a sensible, divided model appropriate to our minds. Maximus says in the *Mystagogy* that "the church made by hands was wisely handed down to us as a guide to something better and an example, through the symbolic diversity of divine things in it."[41] The church enters into diversity as a concession to our weakness, but this diversity is not our proper mode of overcoming the divisions of nature. Our union with the few other human beings who join us in the visible sacred place will one day ascend from image to archetype, from the partial visible gathering to the universal gathering of all things in a cosmic liturgy.

Distance and Its Overcoming

In the year 626, the advance of the Persian army on Constantinople forced Maximus to abandon his monastery at Cyzicus and to go into exile in North Africa. Though at the time he hoped to return quickly to his monastery, the exile lasted for the rest of his life. The physical distance, which separated him from his community, seems to have affected him deeply. His early letters from exile dwell so intently on physical absence that Polycarp Sherwood has seen "nostalgia" in them, and found in Maximus a man "ardently desiring to return to his own monastery, yet prepared to accept his exile."[42] Maximus does not couch his nostalgia in poetic musing, but in a concise philosophical exposition of the nature and power of the letter. His theory treats God as the single and undivided place of the universe, who can overcome the diversity of our particular places through the mediation of the spirit. The letter serves as the material image of this spirit, uniting those of us divided by bodily place to each other in the intelligible place, which is the spirit of God. Since there is presently no English translation of these letters, I here provide all of their relevant content rather than piecemeal quotations.

In a brief letter that contains nothing allowing us to date it to any specific time in Maximus' life, addressed to Stephen in some manuscripts and Auxentius in others, Maximus gently reproves his friend for not writing to him:

> If you are so bold in the spirit that you have decided to neglect letters to your friends, since you are yourself surrounded by an unfailing love which requires

> nothing for its renewal, then I approve your legislation and accept you, who entrust the treasure of that love to the one from whom it arises and in whom it rests. If, however, you think so little of the way we ought to love that you refuse to write out of disdain, I will not cease to do the opposite, since I do not know how to respect my friends for their great concern with the world. If neither of the two causes of your not writing obtains, then receive me, though I am away, and give your conversation to me enthusiastically through letters as nature taught, or, rather, as God in his wisdom exhorted nature. For a word given shape in a letter prods the inactive memory, and wakens it when it slumbers, so that it sees and imprints itself with the faces of loved ones. Through letters, as from a small spark, yearning flares up again like a fire that has been quenched.[43]

The concluding lines of the letter are more than a nice turn of phrase: they introduce an abiding theme in Maximus' letters. Though Maximus thinks of the letter as possible only because of the work of God, the writing of letters is more immediately a work of nature. After all, Christians are not the first to write letters to their friends. Having identified this origin of writing letters, Maximus goes on to specify their character. He does not treat the letter as a means of communicating information. "Conversation" (ἔντευξις) could equally well be translated as "presence," since it literally means "to happen upon," and the derivative meaning of "conversation" seems to have arisen from the fact that, when we run into each other, we get to talking. Taking the term in its root meaning, Maximus' request to Stephen is: "give your presence to me enthusiastically through letters." He shortly explains why the letter should provide presence rather than information. The letter acts as a spur to memory, which otherwise is prone to forget those who are not physically present. Once the memory is spurred by the presence of the words of the letter, it sees, presumably through imagination, and "imprints itself with the faces of loved ones." The letter does not simply unite souls. It reminds the reader of the absent face of the writer. By setting up this tension between presence and absence, the letter inspires "yearning" (πόθος) for the writer of the letter, a theme to be discussed more thoroughly in the letters dated to the early years of Maximus' exile.

Whether the addressee of the above letter is Stephen or not, we have a second letter more clearly attributed to Stephen. From a third letter, which identifies Stephen as both priest and superior, and from the fact that Maximus in the following letter refers to his addressee in the plural, Sherwood concludes that this Stephen may be the head of a monastic community, probably the community at Chrysopolis, where Maximus had spent the first years of his life as a monk. The second letter, at least in that portion of it which survives, devotes itself entirely to the amplification of the themes we have seen earlier, now set in the context of an overcoming of physical distance:

> It is characteristic of spiritual love that it not only benefits those in our presence who need it, but also addresses those who are away. It does not allow souls to be separated along with bodies, and it does not allow the power of the word, through which the soul bears the image of the creator, to be circumscribed by place. It allows those in the presence of loved ones to discuss suitable topics in speech, eye to eye, and those who are apart to converse through letters. For nature, by the grace of God, wisely conceived this way to an inseparable union of those physically separated from each other by a great interval of place. For the word contained in speech and letter introduces to the soul a more enduring, indelible memory, such that, whatever may happen, it may see loved ones as always present in the spirit by love, and be embraced and recalled from all things grievous. Do not, then, pass over me, your child and disciple, reverend fathers, but, since you are both disciples and teachers of love, nourish my hungry soul with a word about the virtues, and illumine my mind, oppressed as it is by the gloom of ignorance. Look to the reward this will store up for you in heaven, and perhaps you will become a little mightier in love and more venerable in instruction. I know that to converse directly with God and to exercise providence compassionately over subordinates are not of equal honor, but even the latter is not alien to those whom God acknowledges.

Again, Maximus notes that, though God gives the letter its power, it is immediately the work of nature. And again, Maximus claims that the role of the letter is not to transmit information, but to reestablish the memory of loved ones in the soul. This is not a mere fantasy that creates a friend to assuage loneliness, but the genuine presence of the faraway friend. It is because Stephen and the other "reverend fathers" can become genuinely present in their letters that Maximus can request from them "a word about the virtues." In this letter, Maximus specifies that love is the means by which his companions become present, and spirit is the medium. We can presently see that this medium has great power, for it allows the soul to escape the limitations of place, understood as a form of "interval" or distance. Maximus' language here recalls the overcoming of the division between heaven and earth in the *Ambigua*, where the sensible creation does not "divide itself into places by any extension at all, because it becomes as light as spirit, and is held down to the earth by no bodily weight."[44] By acquiring the nature of spirit, the sensible creation becomes "one and undivided," and opposes itself to place, which divides and is allied with the earth. In both the *Ambigua* passage and this letter, Maximus identifies place not with gathering, but scattering. The letter, like the church, escapes the divisive activity of the visible place by appealing to the undivided place of the spirit. The spirit in question seems to be one step further into the incorporeal realm from the Proclan cosmic place, which remains a body even though it is made of immaterial light. The spirit of the letter enables the converse of souls, not bodies, and is the realm of "the word."

One further letter may serve as introduction, before we take up the letters that most clearly articulate the epistolary art. Maximus addresses this letter to "Constantine, the sacellarium." We do not know who Constantine is, but the sacellarium is the emperor's treasurer, and so this Constantine may be a friend of Maximus from his days at the imperial court. The letter begins with a reference to the end of some conflict. Combefis, followed by Sherwood, supposes that the peace to which Maximus refers is the victory of the Byzantine emperor Heraclius over the Persian emperor Chosroes in the year 628.[45] I have provided here only the beginning of the letter:

> There arrived, with the peace, a venerable letter of my master, the one whom God protects. Its structure seemed to bear him whole, for solemn words with a suitable structure tend somehow to manifest the disposition of the soul. I was pleased, as though I stood in the presence of my master, and I greeted him in the spirit. Then, having thoroughly embraced him with the expanse of my heart, I put it aside. I glorified Christ our God who wisely implanted the law of love in human beings. Because of it, those who know how to care properly for the seeds of love can never be parted from one another, even if physically separated from one another by a great interval of place.[46]

In addition to repetitions of earlier themes, Maximus adds here the suggestive claim: "solemn words with suitable tropes tend somehow to manifest the disposition of the soul." We have seen above that, in the letter, the soul puts itself in words. Here Maximus suggests what kind of words best serve this project. He relies on one of his favorite distinctions—between "word" (λόγος) and "way" (τρόπος)—to make what we may here call a distinction between content and form. If the letter is to manifest the soul, its content must avoid the trivial in favor of the profound, and its form must suit itself to the manifestation of the soul, presumably by using a style that does not hide the character of the author.

Maximus' greatest single letter devoted to epistolary theory addresses itself to John the Cubicular, a courtier in Constantinople that Maximus may have known from his time at the imperial court:

> Some, arousing a worldly love for one another, preserve it by means of their physical presence, on account of the forgetfulness which naturally withers all

yearning aroused for bodies alone. When they are so disposed to one another, their memory tends to be quenched. Even when this very disposition seems able to preserve their worldly relation to one another, satiety follows and divides them, and the whole preceding disposition is obliterated. Often, even the absence of one of the causes which effect this disposition, or some small circumstantial occasion, makes love aroused in this way revolve quickly into hate.

On the other hand, those who are bound to each other unceasingly with the chain of a godly love when together, simply extend this chain of love when they are apart. They have this love in them because of God, who is love, and who supplies the saints with the power of being able to love. And this is only fair, it seems to me. The disposition of those disposed to one another like liquids tends to drain away, since liquids naturally do this. But the affection of those disposed to each other like solids customarily stands its ground, absolutely unmoved and unshaken, since solids are steadfast and remain always the same. It extends to them a mutual yearning for the unbounded to the extent that God, the author of this relation, draws to himself those who are bound to each other in the spirit. By loving you in this way, you who are so dear, I hold you as inseparable, whatever may happen, and with me in soul, deeply impressed on me in the spirit, even if there is between us a great length of time and a large interval of place. I never cease to see you and greet you in my mind. Believing that you love me equally, if not more, I neither refuse nor hesitate to write what I must, knowing that the two of us have one soul through the spirit. I know that you will not consider yourself scandalized when you hear this, if you think of the fearful descent of God toward human beings, beyond mind and speech. Receive favorably the bearer of the present modest composition, and be all things to him, like the one who gave laws to human beings, that they might assist each other, and make their neighbor their own. They do not consider their neighbor to be another, since their mutually adopted disposition makes them into one another. As the divine word explains, do not merely "rejoice with those rejoicing," but also "mourn with those mourning." Do not simply love, but love your neighbor as yourself. I could even speak of a new commandment which enjoins us to offer our souls themselves in behalf of others for love's sake in time of struggle. The Lord who gave us this command fulfilled it with a deed, not refusing to offer his soul in behalf of ours, an example for us of perfect acceptability. By it, the law of self-love, which from the beginning, through pleasure, has led us by deception away from God and from each other, is slowly quenched. This law, by teaching us to believe in many gods or none at all instead of one, splintered our ability to reason in matters of the flesh. By cutting our single nature into many parts, it inflamed our capacity to rage against each other for pleasure's sake. Be then, as I said, all things to those in need of your assistance, that you may find such a God in you, who consents to be all things to all things on account of his love for humanity.[47]

We have seen that Maximus identifies memory as the faculty of soul stirred up by the letter. Here he demonstrates how different the presence of a loved one in memory is from the presence of the loved one in the body alone. The two differ so strikingly that their loves merit different names. The love of another's body Maximus calls "worldly love," and he suggests that, unlike the love that fosters memory, this love becomes the victim of forgetfulness. Forgetfulness "naturally withers all yearning aroused for bodies alone." A vicious circle is at work here. Forgetfulness withers bodily yearning, and so the lovers try to stay in one another's presence, but this presence produces the very forgetfulness that withers their yearning. Bodily presence, if sought for its own sake, kills memory, because you do not need to imagine your loved one in memory if that loved one is present to you in sensation. Second, and more importantly, memory deals with the soul as its proper object, and so a neglect of the soul in favor of the body allows the memory to atrophy. Bodily presence does not only kill memory: it eventually also kills the yearning for the body that is present. Satiety, as when one gets tired of seeing the same body all the time, or a change of disposition in the body of the beloved—such as illness or aging—"makes love aroused in this way revolve quickly into hate."

Godly love, on the other hand, does not require the physical presence of the loved one. Is it enhanced by physical presence? Maximus does not tell us in this letter, but we will find a clue to his answer in a later one. In this letter, we see only that this love is capable of extension over any distance, like a chain. Maximus here goes further than we have so far seen in describing the source and operation of this love. He says that it comes about "because of God, who is love, and who supplies the saints with the power of being able to love." This love, which comes about in the medium of the spirit, is not merely a gift of God. It has in itself an essential relation to God, presumably because the unity it brings about between friends depends on the prior unity of God. The love of friends for each other "extends to them a mutual yearning for the unbounded to the extent that God, the author of this relation, draws to himself those who are bound to each other in the spirit." We saw earlier that the letter brings about the presence of the friend in the soul, not a mere wishful imagination of that friend. Here Maximus explains how this is possible, with a claim that he suggests will surprise his friend John: "the two of us have one soul through the spirit." He straightaway attempts a theological justification for his claim, in which he suggests that it is borne out by the doctrine of the incarnation.[48] For Christ, too, has one soul with us, since he is, as the apostle Paul says, "all in all."[49] Christ, too, advises a change in our being with the command: "love your neighbor as yourself." This means more than treating your neighbor in the same way that you treat yourself. It means thinking of your neighbor *as* yourself: "they do not consider their neighbor to be another, since their mutually adopted disposition makes them into one another." Where Christ may have been speaking about any human being, Maximus here treats the command as applying specifically to the friend, with whom we have a "mutually adopted disposition." The effect of this unity is shared experience, both in happiness and in sorrow.

At the end of the letter, Maximus rephrases the conflict he posed at its beginning. The love that enables the letter to wield its power is exemplified in the action of Christ, who offers his soul on behalf of ours. This form of love acts in direct opposition to self-love, to which Maximus attributes the characteristics of the worldly love he discussed at the beginning of his letter. Self-love is divisive: it cuts off the self from God and from other human beings, and so makes us capable of "rage against each other." Its emotional manifestation is the search for pleasure as replacement for the consummation of ourselves in union with others and with God, but it also transforms our very being: it cuts "our single nature into many parts." By cutting myself off from others, I also break up our common nature. Maximus suggests that this form of love also has an effect on knowing: it teaches "us to believe in many gods or none at all instead of one." This cryptic claim seems to mean that self-love, by breaking off our inner tie to all other beings, does not allow us to see the unity which binds all things together. We replace that unity either with nothing, in which case we cease to believe in any god at all, or with many gods, in which case each human being becomes her own deity, isolated from the rest. The overcoming of self-love in Christ replaces this division with the original unity of our nature.

We may be tempted to think of a supernatural agency at work here, and Maximus would agree, but the mechanism of this agency should be familiar to us, as he makes clear in a later letter, addressed to Conon. From the fact that Maximus seems to regard Conon as his own superior,[50] Sherwood dates this letter after the departure in 633 of Sophronius, friend of Maximus and future patriarch of Jerusalem, for Alexandria. Maximus here repeats his "one soul" doctrine, and specifies that the effect of this unity is a form of faith, specifically faith in each other:

> I believe what has been handed down and taught to me, that God is love. Since he is one, and never ceases to be one, so he makes one those who live according to his love. Even though they are many, he bestows one heart and soul on them, so that, since they have one soul, they may also know each other's hearts. They do not labor, seeking out of ignorance things that are unclear to them, each one guessing at the disposition of his neighbor to himself. Because of this, reverend father, I believe that, by the grace of God which is moved in you, you are ignorant of nothing in my heart.

> Likewise, even if it is rash to say so, I am ignorant of nothing in yours. I dare to refuse for the moment to set out for you, knowing that you will not believe the refusal to be out of disobedience, but out of a physical suffering which forcibly keeps me from the road, and does not consent that the enthusiasm of my soul emerge in my actions. But though I am physically absent, I can never be parted from you in soul, since I am always present in the spirit through the present letter.[51]

The unity between the writer and the reader of the letter is brought about through belief. Because Conon believes that Maximus reveals himself truly in the letter, and that God does truly provide a link between two creatures so that one may know the thoughts of another, he can believe Maximus' reason for not visiting.

I present one final letter on epistolary theory out of its presumed chronological order, because only here does Maximus address the question of the value of bodily presence even for those who are joined in the spirit. This letter provides an important counterpoint to the overcoming of the visible place that the other letters hope to accomplish in themselves. I provide here only the beginning of the letter, which goes on to describe the forced baptism of the Jews ordered by the Byzantine emperor Heraclius in 632 and places the letter chronologically before the letter to Conon mentioned earlier. In it, Maximus reiterates the superiority of spiritual over physical presence, but he also adds a rare qualification, closing with the hope that he will someday see his addressee face to face:

> The yearning of the flesh tends to be withered by time, departing when those whom it joins together are separated in place. For it acquires its sustenance from sensation, which can in no way captivate those who are not in each other's presence. The yearning of the spirit holds those who are always in each other's presence in the mind, bound together by it, though physically parted from each other. It cannot be bounded in place or time, for it acquires its existence from the mind, which is in no way divided or constrained by the separation of bodies from each other.
>
> I have long been made worthy of possessing this yearning for you, venerable ones. I seem always to see you in my presence and, in conversing with you, I seem to sense you. There is no time or place which can separate me from the memory of you. No matter what happens, memory reveals that you are present spiritually, expelling the stench of my provocations which do not bear the captivating fragrance of your divine grace. I believe that memory does not imagine you in vain, venerable ones, but senses your true presence. When I imagine you, I am made absolutely certain of your presence. For the power which is in you by the grace of God supplies very clear evidence of your presence in its use of memory to expel my troublesome demons. It is not to be marvelled at that the God of signs and wonders makes those

> who are physically absent to be present in the mind, as God himself knows along with those who divinely perform such things by him. Such presence is greater than when bodies are in each other's presence in place. If in memory alone, reverend fathers, you supply me with such an antidote to my irreverent provocations, arriving invisibly in the spirit, by how much more would your presence to my eyes, sanctifying my hearing through living speech by your divine words, and brilliantly instructing me in virtue by your ways?[52]

In an earlier letter, Maximus requested "a word about the virtues" from Stephen and his companions. Here we see the effect he expects. The presence of his friends in his memory destroys his "provocations" (λογισμοί), a term that has a technical sense in monastic literature, and means thoughts that interrupt the monk's pursuit of virtue and contemplation.[53] At the beginning of this letter, Maximus ranks presence in memory higher than presence in place, yet a few lines later he compares the spiritual presence of his friends to their additional physical presence in favor of the latter. If they were in his physical presence, they could sanctify his hearing with "living speech" and, by their "ways of life," they could exemplify for him the path of virtue. This is finally the force of the term "yearning," which Maximus uses repeatedly in his letters. It is the desire for the more complete presence of one who is already present in the spirit. Yearning (πόθος) in this sense is opposed to passion (πάθος), which is a desire brought about only by the absence of its object, and is not able to maintain this dialectic of presence and absence. Because the term "yearning" is prominent only in those letters that Maximus wrote shortly after his exile from Cyzicus, Polycarp Sherwood suggests that we see here Maximus' nostalgia for his lost monastery. This may be true, but Maximus extends the meaning of the term beyond his personal yearning to the common yearning felt by friends for each other's bodily presence, and to their yearning for God. Maximus derives from his personal yearning an expression of a common yearning for presence both in body and soul, in both the visible and divine forms of place.

Whether Maximus thinks of the visible presence of his friends as necessary only because of his orientation toward the visible, or whether he genuinely thinks that visible presence perfects spiritual presence, is difficult to say. His work is permeated with the desire for a spiritual body, one not subject to the divisive effects of visible places. And yet the *Mystagogy* also finds in the church a visible place that brings about unity rather than dispersal, and his letters identify themselves as a means whereby the memory "imprints itself with the faces," not just the souls, "of loved ones." Eriugena will focus more intently on the faults of the visible place, destined to last only as long as the present world, and the virtues of the intelligible place, suited to a spiritual body and identified with paradise. Yet in the margins of his philosophy, Eriugena, too, will find a positive role for the church and for the visible generally.

Eriugena

When scholars describe the role of place in Eriugena, they typically ignore his use of place as a category of sensible things, and focus instead on his more Neoplatonic treatment of place as intelligible.[54] But the term "place" appears in Eriugena more frequently as a sensible than an intelligible category. The sensible use of "place" also seems to represent Eriugena's earliest thinking on place, since it is the only treatment of place to appear in his early work, the *Treatise on Divine Predestination*. In this treatise, Eriugena distinguishes things that are in God from things that are under God.[55] Things that are in God—the content of divine names such as "being," "life," and "intellect"—are only metaphorically in place and time. Things that are under God, like every visible creature, are literally in place and time. Socrates, then, is literally in the marketplace, but his being is only metaphorically in the marketplace, since this being belongs properly only to God, who is not properly in place.

The Visible Place

The sensible category of place has an air of tragedy about it. Eriugena refers to "the dispersal of place and time" so often that it almost becomes a cliché.[56] By being in only one place, I separate myself from the other places, and so perpetuate the dispersal of the visible world into many different places. The *nutritor*, one of the two interlocutors whose conversation constitutes Eriugena's major work, the *Periphyseon*, claims that the "dispersal of place and time" is also responsible for my distinction from other human beings: "the diversity of human beings from each other by which the species of each is distinguished from the others and the mode of stature is varied, does not arise from nature, but from the defect and diversity of places and times, lands, waters, airs, foods, and other like things in which they are born and raised."[57] Place not only separates me from the sight of other beings, but my place is responsible for my acting and looking different from other beings even if we later come together in the same place. Eriugena's mention of "defect" here is crucial, for he, unlike Maximus, includes all of these distinctions under the rubric of the fifth division of nature, the one which results from human sin, and so they have no role to play in the world's ideal state. Maximus, as we have seen, treats only the division of human nature into male and female as the result of sin, even though he does in his letters experience the visible place as a dispersal. Eriugena's more radical claim allows him to treat place itself as inherently sinful. After all, the Aristotelian category of "where" (πoῦ) need not have any positive effect such as gathering and defining the thing in place. The *nutritor* simply calls it "the seat of particular members of the body,"[58] leaving it in a state of near total passivity. The *alumnus*, the *nutritor*'s partner in the conversation of the *Periphyseon*, stops short of considering place and time a punishment for sin, saying that "the order and beauty of the whole visible creature exist only in the variations of things through places and times," and that "they were made more for the education of human nature and its return to its creator than for a punishment of sin."[59] Place and time are merely the result of sin, not its punishment. Their diversity allows harmony—one of the constituents of visible beauty—to come about, since harmony requires a diversity of parts. Place and time also serve to direct human nature at its creator by their own insufficiency. They cannot have caused themselves, and are unsatisfying as ends in themselves, and so they direct us to something beyond them.

Eriugena's most extended treatment of place as a sensible category comes about in the course of explaining Augustine's position on place. Augustine seems to disagree with other authorities in the church, first of all by denying that there can be a place without a body, a view that he seems to have received from the Aristotelian tradition. Augustine's argument, according to Eriugena, is that "if places and times were before the world, they are plainly eternal. And if they are eternal, they are either God or the principal causes of all things constituted in the divine wisdom, which would be most foolish to think."[60] Augustine thinks of place as a form of extension, and so may easily deny that there are "places above heaven," since neither God nor the primordial causes have extension. Augustine also seems to disagree with other authorities in the church by saying that place and time can be separated. He says that God "moves himself without place and time, he moves the created spirit through time without place, and he moves the body through time and place."[61] This is a view that Eriugena himself employs in his early work on predestination, before his reading of Maximus and development of his own theory of place. The *nutritor* and the *alumnus* do not wish to adjudicate between the authorities on these points, and so they leave the matter to the discerning mind of the reader. They elsewhere, however, accept the Augustinian view for the visible place, while at the same time maintaining that there is an intelligible place, which is "above heaven" and is eternal.

Place as a sensible category cannot last beyond the end of the world. At the end of the world, all bodies perish, and if they perish, then place must perish too. Otherwise, we are faced with the paradox of a place without a body to be contained by it. As the *nutritor* says: "when that does not exist which needs to be placed and circumscribed, how will there be place? For if it places nothing, place will not exist. If that which was placed perishes, for what reason will there be place? Place is nothing at all when it does not place anything."[62] After the end of the world, place in the Aristotelian sense survives only as a definition (*ratio*), which remains in the "causes" of the places we now experience.[63] The *nutritor* cautions us, however, that the kind of place that perishes is "not the definition of things which remains always in the soul, but the space by which a quantity of bodies is extended."[64] Only the sensible variety of place perishes with the world, while the intelligible place (the definition of things) is eternal.

The sensible category of place is of limited value in yielding knowledge. Like most of the other Aristotelian categories, it reveals a circumstance of things without defining their substance. The *nutritor* claims we should not try to define the substance of a thing, "since it cannot be defined, but is encompassed, within certain limits, as it were, by its circumstances—I mean place and time, quantity and quality, relation, junction, rest, motion, disposition, and the rest of the accidents by which a substance, unknowable in itself and subsisting without definition in the structure of its subject, is revealed only as to the fact that it is, not as to what it is."[65] The circumstances of a being reveal to the observer that something is there, but they do not tell the observer what it is. If I see something with the qualities of being orange, circular, and citrus-smelling, the quantities of being about four inches in diameter and about a half pound in weight, I know that this something exists, but I do not know what it is. When I call it an "orange," I simply designate the aggregate of perceptible circumstances, not the substance underneath them. The *nutritor* says that we do not understand the "structure of its subject" if we think that the substance of this something can be defined. Its substance lies outside the range of definition. Certainly, I see something and, unless someone intentionally deceives me, I am right to treat it as food. I do not, however, get at the substance of the thing when I treat it as food. Its sensible, useful circumstances do not touch what it is. I ought always to remember that the substance of the simplest thing is hidden from me, just as, for the same reason, I ought to remember that God in himself is hidden from me. The scholar of theology knows this, and so he does not "presume to ask 'what is it?' of the divine substance, since he thinks of it in purity, that it cannot be defined, it is not one of the things that are, and it overcomes everything that can be defined." If the substance of all things lies not only beyond their sensible circumstances, but also beyond intelligible definition, then it is not only the sensible category of place that fails to grasp it. Even the intelligible place—place as definition—does not define the substance of a thing, though it has many advantages over place as a sensible category. It is to this intelligible place that we now turn.

The Intelligible Place

"Place. . .is constituted in the definitions of things which can be defined," the *nutritor* claims at the head of the *Periphyseon*'s major discussion of place.[66] This discussion comes about as the two interlocutors proceed one by one through Aristotle's categories, and find that none of them apply to God, but their account of place here is not Aristotelian. Granted, the *nutritor* does follow up his claim with an explanation that sounds Aristotelian: "place is nothing other than the circumference by which anything is contained by certain bounds." However, this Aristotelian definition is immediately removed from its Aristotelian context. The *nutritor* goes on to say that things encompassed by place may be incorporeal as well as corporeal, already a step beyond Aristotle. He adopts the broader Iamblichan explanation of place, identifying a thing's place with its definition, which is the same for all the visible things that fall under the definition. If I am asked the place of my wallet, I may say my back pocket. An Aristotelian would say the limit of the body that surrounds the wallet, which in this case does more or less correspond to the back pocket, since the two have the same shape. In their discussion here, however, the *nutritor* and *alumnus* would say that the place of the wallet is its definition, no matter what its surrounding body.

Eriugena dismisses the Neoplatonic and Aristotelian doctrine of natural places, at least when place is used in the sense of definition. The *nutritor* says: "if body is one thing and place is another, it follows that place is not a body."[67] It makes no sense, then, to speak of the world's four regions—earth, air, water, and heavenly fire—as natural places for certain sorts of things. The air cannot be the natural place of birds, because the air is a body, and bodies cannot be places. None of the Aristotelian categories are bodies, though some of them produce bodies when combined with other categories. Place is not one of the categories that can produce bodies, no matter what it is combined with. It is one of the categories that "appear in nothing and are always incorporeal."[68] To say that one body is in another as its place is to confuse the category of place with the category of quantity, which *does* become apparent to the senses when combined with other categories. Quantity is first of all "a certain dimension of parts which are separated either by reason alone or by a natural difference," and second, "a rational progression of those things that are extended in natural spaces, width, height, and depth, to certain bounds."[69] The *nutritor*'s first treatment of quantity has much in common with Damascius' concept of number as the limit of extension in multitude. It can remain incorporeal, as number does, in which case its parts are separated "by reason alone." The second form of quantity, however, must appear in bodies, those things that are "extended in natural spaces," and so resembles Damascius' definition of place as the limit to extension in mass. Quantity limits extension in mass by providing "certain bounds." When the ancients, then, speak of the air as the place of birds, what they mean is that the quantity of birds is limited within the quantity of air. The limits of bodies are other bodies: these limits are the only residue of the theory of place from Aristotle's *Physics*, which the *nutritor* denies here should even be described with the name of "place," though as we have already seen, both interlocutors elsewhere make liberal use of "place" loosely as an Aristotelian category. In their present discussion, "place" indicates only the limit of the form, not the body, and so it remains incorporeal, having only a metaphoric relationship with the sensible category.[70] Incorporeal things also have places, as examples of which the *nutritor* lists off the seven liberal disciplines—grammar, rhetoric, dialectic, arithmetic, geometry, music, and astronomy—each with its own definition. These disciplines are not just random examples of incorporeal things that have places. The definition of each art contains within itself "innumerable others," so that the liberal disciplines may be conceived as the broad genera under which all other definitions are grouped. The *alumnus* shortly assumes this stronger claim, that "every definition is in a discipline," as part of his growing understanding that every definition must be nowhere else than in the soul, the place of the liberal arts.[71]

The *nutritor* and the *alumnus* go back and forth over the course of the *Periphyseon* on whether definitions have bodies associated with them. After all, when considering the sensible category of place, they conclude that "place is nothing at all when it does not place anything."[72] If there is no body, there is no place. Does this hold for intelligible as well as visible places? The two interlocutors never address this question directly, but for most of the *Periphyseon* they agree that even beings that do not possess bodies subject to the sensible category of place nevertheless possess "spiritual bodies." With this distinction, they make creative use of the apostle Paul's distinction between our present human body, which "was sown an animal body," and our body in the life to come, which "will be raised a spiritual body."[73] Their rationale for the spiritual body is this: a human being is a whole, whose parts are body and soul. If we never stop being human beings, then we do not stop being both body and soul, even if the soul is at times separated from control of the body.[74] When the human being dies, its body is dispersed among the elements. What was air in it returns to the rest of the air; what was earth in it returns to the earth, and so on. In the final destruction of the world, the elements will be totally dispersed into one another.[75] The earth that was my particular body will mingle completely with the earth that was every other body. When I come to control my body again, freed from the results of sin, my soul will choose those parts of the body that are most like itself, the most spiritual parts. These "are very light and unimpeded by the weight of heaviness or baseness. With no intervening delay they immediately arrive wherever the soul commands, just like vision and hearing. No proper philosopher could deny that these parts of the body are drawn from fire and air."[76] The soul, then, chooses fire and air for its new body, since they, in sight, hearing, and smell,[77] are not contained within limits of dense corporeality, but resemble the intelligible character of the soul itself.

The angels presently have such bodies. The *nutritor* tells his *alumnus*: "neither the angelic spirits nor their spiritual bodies, which subsist causally in their spirits, are contained at all within the walls of the corporeal creature."[78] Where Plotinus suggests that all bodies are contained in their souls, Eriugena maintains it only for angelic bodies. Unlike humans, whose animal bodies are truly confined within the limit of the body that surrounds them, the angels possess "spiritual bodies united to their intellects." We may be tempted to think that the term 'body' here is little more than a figure of speech, but the *nutritor* hastens to assure his *alumnus* that "they appear not imaginarily, but truly. There can be no doubt that true and spiritual bodies proceed from true reasons." Although their bodies are not limited to visible places and times, the angels confine themselves in visible places and time when they appear to human beings: "they transform their spiritual and invisible bodies into visible apparitions in order to reveal themselves in space and time to the mortal senses," and "they accept this accident not for their own sakes, but for the sake of those men of whom they are in charge."[79] Christ, too, when he ascended into heaven after his resurrection, did not move to a different visible place. He simply "hid himself in the subtlety of the spiritual body, unseen by the fleshly eyes of the apostles."[80] His resurrected body is a spiritual body that, like the angelic body, has no visible place. The *nutritor* asserts: "they are worthily to be refuted who attempt to add the body of the Lord after the resurrection to some part of the world."[81] Our bodies, too, will leave behind place as a sensible category in the resurrection. We will "be made equal with the angels," by which he means that our bodies will be "changed into a heavenly quality." These bodies "put down everything earthly, cannot be grasped by mortal senses, and are free from all the circumscription of place."[82] Such bodies will not exist independently of their souls, as bodies do now. The *nutritor* says: "inferiors are naturally drawn to their superiors, and are absorbed by them, not so that they cease to exist, but so that they may be preserved in them more."[83] The body will be absorbed in the soul, though not in such a way that it is reduced to soul. In this absorbing preservation, we see Eriugena's version of what Maximus calls "unconfused union," where the parts are preserved in their existence even though they do not appear independently of the whole. Eriugena seems to have derived this idea directly from Maximus, for he immediately explains it with Maximus' example of air and light: "air does not lose its substance when it is wholly turned into the light of the sun, although nothing appears in it but light."

The *nutritor* and his *alumnus* do not always smile so benevolently on the body. In one particularly severe passage, the *alumnus* recalls with horror the Augustinian doctrine that our bodies in the resurrection will be spatially extended, the same height as our present bodies, and differentiated into male and female. To put the best spin possible on Augustine and his peers, the *alumnus* says: "to these great and divine men, who were mindful of the timid thoughts of the simple faithful people, it seemed more useful to preach that earthly and sensible bodies will change into heavenly and spiritual bodies than that bodies and all bodily and sensible things will not exist at all."[84] Augustine does not want to scandalize the simple-minded, and so he describes a heaven filled with bodies that look roughly the same as they do now. The truth as the *alumnus* describes it here is that "bodies and all bodily and sensible things will not exist at all." It is not clear what the *alumnus* means by this. Given the context, he may simply mean that death does away with all bodies composed of the elements in their present manner, while spiritual bodies still come into play in the resurrection. On the other hand, the *alumnus* may have gotten carried away and now wishes to eliminate bodies altogether, even spiritual ones, from the life to come. Whatever the case, the place of the life to come remains intelligible, whether spiritual bodies belong there or not.

Eriugena follows the classical tradition of identifying the pastoral landscape as indicative of a certain state of soul, and he follows the Christian tradition of identifying the paradise of the Hebrew scriptures with such a pastoral landscape. Epiphanius, the fourth-century bishop of Constantia, has said that paradise is "a certain sensible place in the eastern parts of the world, with sensible trees and rivers and everything else which is believed about paradise in the simple manner of the flesh by those who cling to the bodily senses."[85] The *nutritor* roundly criticizes this interpretation of paradise, supporting himself with the authority of Ambrose and Origen. He tells the *alumnus* that "true reason laughs" at Epiphanius' interpretation.[86] Anyone who reads Ambrose knows that "paradise is not a certain forested, local or earthly place, but a spiritual place sown with the seeds of the virtues and planted in human nature. To put it more plainly, it is nothing other than the human substance itself."[87] Where is this intelligible paradise? The *nutritor* says that God planted human nature "in the delights of eternal happiness. And where had he planted it? In that beginning, of course, in which God made the heaven and the earth."[88] Paradise, the intelligible place that is human nature, is itself in the place which is God. It stands midway between visible places and God, who is beyond even the intelligible place. The sensuous language used by the book of Genesis to describe paradise is to be taken symbolically, then, and not literally.

The Cosmos as Sacred Place

When Eriugena relocates paradise to the realm of thought, he removes one more possibility of a sacred place from the realm of the visible. We may well wonder whether our only opportunity to enter a sacred place now lies in the paradise of thought which is our own essential nature. At the very least, Eriugena retains the general sacred character that the Neoplatonists give to the visible world on the basis of its form. The Neoplatonic visible world is an image of the universal intellect, and so is tied to the divine in as much as it maintains a likeness to that intellect. We may say that the visible world has a general sacred character in Eriugena for two reasons. First, everything in the visible world makes itself visible only through its characteristics and, as we have seen, Eriugena rigidly separates these characteristics from the substance of the thing. When I see a thing that is orange, circular, and citrus-smelling, I know that something is there and I call it an orange, but I still know nothing of the substance of an orange, since "a substance, unknowable in itself and subsisting without definition in the structure of its subject, is revealed as to its being alone, not as to what it is." The divine substance is no more knowable than a created substance. But in the case of the divine substance, there are no sensible circumstances that could directly reveal it "as to its being alone." Instead, the circumstances belonging to visible things perform the double role of directly revealing their own substance, and indirectly revealing the divine substance. Other philosophers, like Maximus the Confessor, argue that the "caused" character of created substances reveals the substance of God as what caused them, without saying anything about him other than that he exists. Eriugena implicitly suggests that created substances reveal God not only as this efficient cause, but also as formal cause. Their unknowable being is the image of his unknowable being. When we confront the unknowability of the created substance, we also get a sense of the divine. Eriugena does not suggest that our sight of the unknowable divine substance through the unknowable created substance should be a source of extraordinary experience for us, or that the unknowability of the created substance should change our approach to the creature. The fact that I do not know the substance of the orange does not mean that I should not eat it for breakfast, since I interact with the orange only through its circumstances, which I do know, and which have concrete uses for me.

My microcosmic character as a human being allows me to see the world as sacred in a second way: by overcoming the divisions of nature so that God appears in all nature, both intelligible and sensible. Eriugena acknowledges that we speak about distinct sensible and intelligible worlds, because we can contemplate things both as causes and effects. In his commentary on Genesis 1:5—"and there was made evening and morning, one day"—he unites the visible and the intelligible in a single world, "one day," based on the identity of cause and effect in the things themselves. God is in both: "he is created by himself in the primordial causes.. . .Then, descending from the primordial causes. . .he comes to be in their effects and openly appears in his theophanies."[89] Eriugena hopes not to scandalize us by saying that God comes to be in his effects. He elsewhere explains: "the divine nature is said to come about. . .because what is invisible in itself appears in all the things that are."[90] The substance of God does not enter into the effects, but he "appears" in those effects: they reveal themselves as his circumstances. So far, Eriugena has suggested that God proceeds only into the imagination of the interior sense of the soul, for this is the first place where things appear as distinct from one another, and not as one in their causes. But Eriugena goes on to say explicitly that God extends himself as far as the bodily. He "proceeds through the multiple forms of his effects all the way to the last order of all nature, by which bodies are contained."[91] Using our own capacity for differentiation, we may see that the world before our eyes is nothing but God, not God in his substance but God as manifest to us. Because it sees the visible world as God's own circumstances, this way of viewing the visible as generally sacred differs from the first way, which sees visible circumstances as belonging directly to visible things, and only indirectly to God. Again, though, Eriugena does not think of such a world as changing the structure of our experience. He does not, for instance, describe how our actions in the world would change if we saw the world as God, and he does not say what the shape of such a world would look like. His treatment of the world as God precedes any experience of the world, whether this be the experience of the unknowable or the employment of useful things.

The Church as Sacred Place

Iamblichan Neoplatonism develops the idea of the sacred place with exceptional power because it treats certain places within the visible world as symbolic, and so more directly linked to the divine than most places, which are merely images of forms in the divine mind. The Frankish kingdom of Charles the Bald, where Eriugena lived and worked, had many such sacred places—its churches and cathedrals—but Eriugena's work gives us little chance to see what he thought of them. Like his Christian predecessors, he has no use for places in which the lowest kind of the Iamblichan rite would be performed. The mere performance of bodily actions does not incarnate the divine and cannot lead to salvation. Eriugena says this most vividly in his commentary on the Gospel of John, where he interprets Christ's encounter with the Samaritan woman at Jacob's well.[92] The Samaritan woman recounts the controversy between her people and the Jews: "our fathers worshipped on this mountain, and you say that in Jerusalem is the place where there ought to be worship." Christ responds that the time is coming when God will be worshipped neither on mountains nor in Jerusalem, for "God is a spirit, and those who worship him ought to worship in spirit and truth." Eriugena interprets the passage to mean that the mind and not a visible church is the proper place for the worship of God: "if God were bodily or a body, he would perhaps seek bodily places for his adoration. Now since he is a spirit, he seeks those who adore him in their spirit and in their intellect, through true cognition."[93] A little later, he revisits briefly the theme of the mind as the true temple, paraphrasing the words of Christ: "true adorers will adore me and my father not on this mountain, nor in Jerusalem, but within, in the innermost temple of their heart and their intelligence."[94] Eriugena, like Dionysius and Maximus, rejects a rite that saves bodies without reference to any elevation of the mind. Like Maximus, he prefers to speak of the human mind as the true temple, of which the visible temple is a useful image. Whether the visible temple is merely an image or has some sacred character in itself, Eriugena does not say here.

We do find one other discussion of the temple image in Eriugena. At the end of his earlier work, the *Periphyseon*, he provides his only sustained reference to the temple in all of his writings. It functions for him much as it did for Maximus. The sacredness of the temple is in reality the sacredness of the mind, the true temple, or "church not made by hands," as Maximus and the Apostle Paul put it. Eriugena's temple has three major divisions: (1) the porticos, one for the commerce and prayer of "all the nations of the whole world," and one for the priests and Levites; (2) the outer temple, accessible only to the purified priests; and (3) the holy of holies, entered only by the high priest, and containing the ark of the covenant together with the altars of incense and sacrifice. The reference to the two altars is his only reference to the visible altar in the entirety of the *Periphyseon*. These three divisions illustrate for Eriugena the different states of those who enter human nature, which, as we have already seen, is paradise. He says that "everyone is contained within the bounds of the natural paradise, as though within a kind of temple, each in his own order. But only those sanctified in Christ will enter the interior parts, and only those who are in the high priest—Christ, I mean—and have been made one with him and in him will be taken into the Holy of Holies, the interior part of the interior parts, where Christ is."[95] If the innermost parts of the metaphorical temple are closer to God, are the innermost parts of the literal temple closer to God? Eriugena does not say in the *Periphyseon*, but in a brief passage from his *Expositions on the Heavenly Hierarchy*, he suggests that the eucharist celebrated at the altar is not just one sacred thing in a generally sacred universe. He says, "this visible eucharist, which the priests of the church bring about on the altar from the sensible material of bread and wine. . .is a figural likeness of spiritual participation in Jesus, which we taste with the intellect alone in faith."[96] So far, we see nothing that would distinguish the altar and its contents from other places and things. But Eriugena goes on to call the eucharist the "maximum figure of such participation."[97] It has an exceptionally potent figural character, and we may conjecture that the altar, and the church that surrounds it, also possess this character. If this is the case, we may say that Eriugena, like Maximus, retains a visible rite that elevates the laity from the visible to the intelligible.[98]

Eriugena does not explain in the *Expositiones* why the eucharist should be the maximum figure of participation in Jesus, when God may already be seen in every created thing. The *Periphyseon*, however, does explain at least why we do not typically see God in everything. The *nutritor* says that sin caused human nature to be "thrust down from the height of the spiritual life and knowledge of the most clear wisdom into the deepest darkness of ignorance." As a result, "no one unless illuminated by divine grace and rapt with Paul into the height of the divine mysteries is able to see how God is all in all with the vision of true intelligence."[99] Although human nature is naturally suited to overcome the division between God and creation, sin has made this impossible for human beings now without the intervention of divine grace. The *nutritor* does not say here how this grace is given, and he does not say whether it is given often. He mentions only the extraordinary example of the apostle Paul, who describes a case of rapture to the third heaven in his second letter to the Corinthians.[100] We might assume a liturgical context for Paul's rapture from Eriugena's location of it in the "divine mysteries," but we cannot conclude confidently that Eriugena, like Maximus, considers the visible church to be the place where we now overcome the divisions of nature.

Abbot Suger

Abbot Suger, born in 1081, was not an early medieval Neoplatonist. His renovation of the abbey church at Saint-Denis, however, has invited a scholarly debate over the effect of the early medieval Neoplatonists on Gothic architecture, and so a consideration of his renovations is a fitting *envoi* to the period. Erwin Panofsky's groundbreaking work on Suger sets his renovations firmly in the tradition of Dionysius and his translator, Eriugena.

Panofsky is confident that Suger read Eriugena's translation of Dionysius on the basis of several passages in Suger's account of his own work, *The Book of Suger, Abbot of St.-Denis, On What Was Done Under His Administration.* The first is Suger's reflection on the cross of St. Eloy and Charlemagne's crest, both placed on an altar he had paneled with gold. The beauty of the cross and crest, both studded with gems, prompts Suger to lose himself in meditation. He tells us that this "upright meditation persuades me to dwell on the diversity of the holy virtues, by crossing from material to immaterial things."[101] Elevation from visible beauty to the beauty of virtue has a long tradition in Platonism, beginning with Plato's *Symposium*, and is common to both the Augustinian and Dionysian traditions. We could perhaps pinpoint Suger's inspiration if he had said something about the necessity of material mediation, the nature of the mediation performed by the material thing, or the goal of the transformative process. All he has left us, however, is a Neoplatonic commonplace. He adds only that, when the process is finished, "I seem to see myself tarrying as though in some outer district of the world, which is neither wholly in the dung of the world nor wholly in the purity of heaven. God grants that I am able to cross from this lesser part to that greater part by an anagogical mode."[102] Suger finds himself in a world that is neither wholly material nor wholly immaterial. He has not left the body behind, but he does not experience the body in its ordinary capacities. There is some suggestion of a dematerialization of body here, which would elevate it from the "dung of the world" without eliminating all of its bodily characteristics, but the passage is too brief to allow a solid conclusion.

While this first passage seems to rely on the Dionysian ascent from material to immaterial, Panofsky's latter two passages rely on Dionysius' so-called "metaphysics of light."[103] Both are in poems, one inscribed on the panel commemorating the consecration of the church, the other inscribed on the doors. Perhaps the most famous renovation accomplished by Suger at Saint-Denis was the enlargement of the upper choir area, which allowed more light into the church. It is this that he chose to emphasize in the verses he added to the panel of consecration. Suger's verses remind the viewer of a general principle: "that illuminates which is illuminatedly joined with the illuminated, / and that work illuminates which a new light pervades."[104] Panofsky cannot believe that the repetitions of the verb "illuminate" (*clarere*) are only intended to drive home the literal illumination of the church interior, an illumination which would be obvious to the spectator. Instead, he finds that the repetition of the words "almost hypnotizes the mind into the search for a significance hidden beneath their purely perceptual implications." The repetition is "metaphysically meaningful" because Eriugena translated with "illumination" (*claritas*) the Dionysian term for "the radiance or splendor emanating from the 'Father of the lights,'" which Suger could have found in *On the Heavenly Hierarchy*. There, Dionysius explains how visible symbols illuminate us and draw us back from the visible to the intelligible and beyond. Suger, in Panofsky's reading, treats the light that illuminates the choir as symbolic of intelligible light. Panofsky sees the same double meaning of illumination at work in Suger's inscription on the golden doors to his church. Suger says of these doors: "the noble work illuminates, but the work which nobly illuminates / enlightens minds, so that they may pass through the true lights / to the true light, where Christ is the true door."[105] The visible illumination of the door provokes the viewer to pass through its material character as a door to arrive at the immaterial realm of "true light," where Christ is. As Suger puts it, "the dull mind rises to the true light through material lights / and, having once been submerged, it resurfaces when it sees this light." Visible light, then, plays an essential role in lifting the "dull mind."

Panofsky's Dionysian reading of Suger has recently come under attack from a number of scholars, among them Grover Zinn Jr. (who reads Suger as a Victorine), Bernard McGinn (who reads Suger as an Augustinian), and Peter Kidson (who finds no philosophy at all in Suger).[106] After reminding us that the eight medallions on the golden door depict scenes from Christ's passion, resurrection, and ascension, Zinn suggests that these medallions are the key to interpreting the poem inscribed on the door. After the lines which read "the work / should brighten the minds, so that they may travel, through the true lights, / to the true light where Christ is the true door," the poem continues: "in what manner it be inherent in this world the golden door defines." Zinn interprets this last line as explaining how Christ may be present in a material thing, such as the door itself. The eight medallions on the door tell us that the material world only reveals the immaterial world through the historical mediation of Christ in his death and resurrection. The material world without the historical mediation of Christ is only an aggregate of natural forms. Christ, by entering into history, allows material reality to become symbolic of a higher reality. Zinn concludes: "only through the works of redemption associated with Christ can the works of creation, the material world, be rightly understood and interpreted as lamps that lead to the True Light."[107] Though Suger himself never explicitly distinguishes works of redemption from works of creation, Zinn feels confident that this distinction can be attributed to Suger on the basis of its presence in the works of Suger's contemporary, Hugh of St. Victor. Zinn believes the "iconographical program" of Suger depends "fundamentally" on the work of Hugh. Zinn ignores the highly developed symbolic sense of interpretation in Dionysius himself in favor of a Dionysius who thinks that "the material world, by virtue of its created existence, is truly able to point in a symbolic sense beyond itself to the invisible realities of the divine world and

God."[108] This interpretive position is more characteristic of Plotinus than Dionysius, since Dionysius maintains a sharp distinction between symbols and natural forms in the material world. In one respect, Zinn has noticed an important difference between Suger's door and anything in the work of Dionysius. Dionysius finds the power of Christ to be manifested in his incarnation, his manifestation in the scriptures, and his establishment of the church, but he does not reflect on the "centrality of the suffering and humility of Christ," as Zinn suggests Hugh does. Suger's door, then, does make more central the interpretation of history, and specifically salvation history, that governs the Augustinian and Victorine interpretive tradition.

While Zinn's reference to the Augustinian and Victorine traditions of figural interpretation may provide the best explanation of the golden door at Saint-Denis, these traditions do not account for Suger's comments on the elevating power of gems, which make no reference to Christ or to history. Zinn attempts to find a possible resonance in the work of Richard of St. Victor, written twenty years after Suger's work, but Zinn nowhere explains why we should not agree with Panofsky that Suger's method here is Dionysian rather than Victorine.[109] Where Suger mentions a stained glass window, "urging us onward from the material to the immaterial,"[110] Zinn again tries to find a Victorine parallel so as to free Suger from direct dependence on Dionysius and Eriugena. He notes that Suger here employs the Latin term *excitans* ("urging us onward") in a usage found not in Eriugena but in Hugh of St. Victor. True as this is, it does not suggest that Suger here employs a method of historical interpretation, but that the Victorines themselves shared the Dionysian mode of interpretation, and that Suger simply has acquired Dionysius through the Victorines, rather than, or in addition to, Eriugena.

Bernard McGinn's reading of Suger as an Augustinian is equally tendentious, relying on what he himself says cannot be found in Augustine's own words. His interpretation, which refers in detail neither to Suger's text nor Augustine's, makes more sense as a general Augustinian aesthetic than an interpretation of either author. He explains the principle of this aesthetic as "taking words from the ordinary language—such as 'bright'—and revealing their meaning by a rhetoric of *repetitio* while building the plane of meaning up from the mundane to the eschatological."[111] The repetition of the term "brightness" foregrounds it as the locus of meaning in the passages where it occurs. Used literally at first, it subtly changes its meaning as its context shifts from the visible realm to the kingdom which is to come. McGinn's claim here is puzzling, since the passages from Suger we have examined earlier contain no explicit eschatological component. Suger seems to advise simple elevation from the material to the immaterial, from visible light to the light of Christ. If these are eschatological aims, they are so thoroughly realized in the present act of contemplation that they cease to have any explicit reference to a future life. McGinn suggests also that, in Augustine, the central role of beauty is not the revelation of God through the use of light (as in the Dionysian "metaphysics of light" referred to by Panofsky), but the revelation of the harmony of creation.[112] The classical definition of beauty involves the two components of bright color and good proportion; McGinn's Augustine plays up the latter, while his Dionysius plays up the former. He finds Suger's Augustinian side at play in his claim to have equalized the central nave of the church at Saint-Denis "by means of geometrical and arithmetical instruments," so that what "we proposed to carry out had been designed with perspicacious order."[113] Suger does not renovate his church primarily with an eye to elevating the worshipper to the intelligible realm, but to revealing the order that can be manifested in the visible realm, when it is reshaped according to geometrical principles. McGinn concludes that we find in Suger "essentially an Augustinian aesthetic, though with some interesting Dionysian highlights."[114] While he deserves credit for bringing up Suger's Augustinian milieu, McGinn's conclusion seems no less tendentious than Panofsky's, relying as it does on assumptions that cannot be grounded in Suger's own text.

The real question to ask here is: what does Suger think is the elevating principle in light and beauty? Plotinus, as we have seen, identifies form as the cause of the beautiful, but he suggests that form may manifest itself beautifully either as light or as proportion. When form is manifest as proportion, it appears in something properly material, whose parts can be separated from each other, and so set up the ratios necessary to establish a proportion. When form is manifest as light, it appears more perfectly in bodies that are less material. These will be bodies made not of earth, but of air, water, or fire. Plotinus never recommends the construction of such dematerialized bodies, presumably since this would lower the sage's activity from intellectual contemplation to action in bodies, even if these bodies are dematerialized. Then, too, for Plotinus, the earthy body has its place in the order of reality, and its own beauty. To destroy material bodies in the course of constructing immaterial bodies would be to destroy a form in order to reveal a form, something Plotinus *never* recommends.[115] Iamblichus interprets several activities of the Hellenic rites as dematerializing, but he shows an interest in this dematerialization only in the ritual context. The fire of the sacrifice, for example, dematerializes the body of the animal, but Iamblichus does not suggest that we set more things on fire so as to dematerialize more of the cosmos. Dematerialization is valuable only in a ritual context, where the theurgist who contemplates it can bring about the dematerialization of his soul.[116] Light, too, is introduced into the sacred place not so as to dematerialize it, but so as to provide a medium for prophecy. The immaterial light is sensitive to the will of the immaterial gods, and signifies that will in the patterns it makes when reflected onto the walls from a pool of water.[117]

When Suger inaugurates the Gothic period in architecture by redesigning the choir and letting in more light, he produces a space in which the transparent medium of air can fill with light, making the whole interior of the church a place of dematerialization. Vincent Scully recognizes this dimension of the Gothic church, and claims that it is the purpose of the use of light there. The light of Gothic churches serves "to transcend the statics of the building masses, the realities of this world."[118] Whether Suger intends this effect of light is less clear. He claims only that the light provides "illumination," and does not specify whether he means a material or an intelligible illumination. He uses the term "illumination" again when he describes the golden doors, and here he makes clearer what he intends to be the effect of brilliance on the doors. The "material light" of the doors raises the mind of the person who sees them to the "true light, where Christ is the true door." That Suger would take visible light in both the choir and the doors to be the presence of an intelligible characteristic may be assumed, given the common understanding of light at his time. But his reference to Christ as the truth of which the door is only an image makes it less likely that the dematerialization of bodies is his primary purpose in the introduction of light and gold into his church. The dematerialization of the door may be a symbol of the dematerialization of the mind, while the door itself is a symbol of Christ, the "true light" and the "true door." The dematerializing effect of light within the church is thus incorporated into a symbolic context outside of which it has no value, like the light and fire in the Hellenic rites described by Iamblichus.

On the other hand, the cross of St. Eloy and Charlemagne's crest seem to possess the power of elevation to the immaterial not through any symbolic character, but directly, through their own beauty. This is not the beauty of proportion. The beauty of gold is, as we have seen in Suger's golden doors, the beauty of light, an association made also in Plotinus.[119] Suger says little about the gems that adorn the cross and crest, but we may assume that their beauty, too, comes from their radiance, since they have little in the way of proportion, their surfaces typically being rounded rather than cut into facets. Like stillness and light in Plotinus, the beauty of these bodies seems to derive from the direct presence of an intelligible characteristic, made possible by the effacement of their material characteristics. Surrounded by these dematerialized bodies, Suger finds himself in a dematerialized place—"some outer district of the world"—meaning the district of the celestial bodies, themselves composed of the least material element of fire, and thereby freed from the "dung of the world." This treatment of the sacred place is not Dionysian, as Panofsky would have it, since the gold and gems do not serve as symbols, but directly reveal the intelligible. If it is to be associated with a historical tradition, it should be the Neoplatonic tradition of seeking out intelligible characteristics in the landscape, though in Suger's case the exterior landscape has given way to an interior space shaped by human hands.

Yet, even in the case of the gems on the cross and crest, we cannot confidently say that Suger has abandoned the symbolic character of sacred things in favor of a more direct revelation of the intelligible through their dematerialization. Suger tells us that his contemplation of the gems leads him not merely to the intelligible, but specifically to contemplation of the virtues. The mythic context common to both ancients and medievals had established certain gems as symbolic of certain virtues. Suger may expect his reader to understand that his contemplation of the gems relies on their symbolic character as well as their radiance, without his having to make that character explicit. Very likely we find in Suger both the symbolic theology of Dionysius and the Victorines, as well as a new tendency to create the dematerialized place that earlier Neoplatonists were content to discover in the landscape, and to make that place identical with the sacred place. Like Maximus and Eriugena, his predecessors in the Dionysian tradition, Suger presents, almost certainly without knowing it, a tension between the symbolic character of the sacred and the sometimes immaterial character of the place.

CHAPTER 5

THE LOSS OF PLACE

When Heidegger writes of the sacred place during the National Socialist period in Germany, he evinces a nostalgia for an all-encompassing world of myths, a nostalgia seemingly also felt by the Hellenic Neoplatonists when they try to provide a philosophical groundwork for the Hellenic rites, even as those rites are being slowly extinguished by the Christians. But also during the National Socialist regime, and more especially after its demise, Heidegger writes of a sacred place that does not make the gods present through a world of myth, but allows them to remain absent even as they are gathered into the place. In the gathering of the divinities into the place, human beings "await the divinities. . .and do not mistake the signs of their absence."[1] Heidegger's conception of the sacred here resembles that of the ancient Greeks much less than it does the conception of certain late medieval mystics and above all Meister Eckhart (*ca* 1260–1327). In Eckhart, we find a God whose truest expression is absence, who lives without reason, and who imparts these characteristics also to creation in its ultimate ground.

Heidegger's connection to Eckhart has been identified by Laszlo Versényi and more thoroughly developed by John Caputo.[2] Versényi quotes a sermon of Eckhart's entitled "The Innermost Ground," in which Eckhart says, "life lives out of its own ground and springs forth out of its own: therefore it lives without why, simply living itself." This same language finds its way into a poem by the seventeenth-century pseudonymous writer Angelus Silesius, entitled "Without Why." The poem reads in full: "the rose is without why; it blooms because it blooms; / it pays no attention to itself; asks not whether it is seen." It is Silesius' description of living without why that Heidegger picks up on explicitly in his 1957 lecture course, "The Principle of Reason," in which he brings up Silesius and his poem as a foil for Gottfried Wilhelm von Leibniz and his "principle of sufficient reason" (*principium reddendae rationis*). The principle of sufficient reason claims that *everything* has a "why." Everything about the blooming of the rose can be explained by reference to something other than the bloom itself. We may say, for example, that the rose blooms in order to attract bees. Silesius' poem suggests just the opposite, that the purpose of the rose's bloom is nothing other than its own existence. It "blooms because it blooms." Heidegger explains how Leibniz and Silesius can both be right, by showing how each approaches the rose from a different perspective. Leibniz' principle of sufficient reason "holds *in the case of* the rose, but not for the rose; in the case of the rose, insofar as it is the object of our cognition; not for the rose, insofar as this rose stands alone, simply is a rose."[3] Leibniz approaches the rose from the perspective of human knowing. Insofar as the rose is an object of knowledge, we can find an exterior purpose for its every action. Nothing about it is for its own sake. But if we cease to look at the rose in relation to us, and approach it from its own perspective, as it were, these purposes no longer hold. It is "without why." Caputo explains that Silesius has here opened up for Heidegger "access to a region where there are no 'objects' but only 'things' (*Dinge*), which are left to 'stand' (*stehen*), not 'before' (*gegen*) a subject, but 'in themselves' (*in sich selber*)."[4] Take the example of the jug from Heidegger's essay on "The Thing." We can explain how the wine came to be in the jug by describing the cultivation of grapes and the fermentation process, and we can explain that the wine is in the jug in order to satisfy thirst. Heidegger prefers to explain the wine not with reference to mechanical causes or purposes, but with reference to what we may now call its grounds: the earth, sky, mortals, and immortals. These four are not outside the wine: they stay within it, and it is the jug, as Caputo says, that "collects 'the four' together. Thus it represents the essential features of Heidegger's interpretation of *logos* as that which collects together."[5] For our purposes, this is the essential feature of Versényi and Caputo's archaeology of Heidegger's dependence on Eckhart. The Eckhartian conception of living "without why" has for Heidegger a connection to the role of place as gathering.

The immortals that are gathered in the place do not become present. We cannot gather Bacchus into the wine through consecration because we lack the mythic context that could allow such consecration. The immortals gathered into our wine must remain unnamed, outside the context that could make them present to us. In this, too, Heidegger probably unintentionally follows a trail laid out by Eckhart. In a sermon dedicated to the claim that "all things that are alike love one another," Eckhart explains the nature of our union with God, and the nature of the God we seek. Eckhart explains here, as he often does, that we become united with God only by an abandonment of ourselves as creatures. He says: "God withholds nothing of his being or his nature or his entire divinity, but he must pour it all fruitfully into the man who has abandoned himself for God."[6] This abandonment is not of certain self-interested actions or emotions. It is specifically an abandonment of oneself as a created being, with the finitude and accessibility to understanding implied in the nature of creation. Eckhart says later in the sermon that not all of oneself is created. He speaks of "a light that is uncreated and not capable of creation and that is in the soul." It is this light that both "comprehends God without a medium" and acquires "unity with God." Eckhart's language here makes it sound as though this light will see God as present with a greater capacity than any of our created powers. The power of sight sees only the icon of Christ; perhaps the uncreated light could see Christ himself. But Eckhart shortly says that this light "is not content with the Father or the Son or the Holy Spirit." It is not even content with the single essence in which they come together as one God. Instead, it "wants to know the source of this essence, it wants to go into the simple ground, into the quiet desert, into which distinction never gazed, not the Father, nor the Son, nor the Holy Spirit." The Father, Son, and the Holy Spirit can be named. They enter into the ritual and mythic context of the church, and so become present in its various acts of consecration. Our uncreated light wants a divinity that can only be characterized with the language of emptiness and absence. It is "a quiet desert," "where no one dwells,"—"a simple silence." This divinity cannot enter the ritual and mythic contexts that would allow it to become present in language and visibly grounded in the icon. It is united only to the uncreated light of the soul and shares the unknowable character of that light.

While such a conception of the divinity has precedents in earlier medieval authors, it is only at the end of the Middle Ages that the unknown God is set so radically beyond the God who becomes present within the walls of the sacred place, with equally radical consequences for philosophical reflection on both place and sacred place. A multitude of late medieval and Renaissance authors contribute to the shift from place to space and to the change in the nature of the sacred place. In examining the Neoplatonic contribution to the loss of place as a topic for philosophy and its consequences for our present-day reflection on the world, I will focus solely on Nicholas of Cusa (1401–64), not because he is the only one to speak about place in the Renaissance, but because he is profoundly influenced by the early medieval authors we have surveyed, and he formulates his reflections on place in terms that remain relevant for us. In Cusa's work, the elevation of God outside a ritual context ameliorates the brutal strife that can occur when different historical peoples clash, but it also degrades the sacred place as the place where all things are united, and opens the door to a more arcane form of union.

Homogeneous Space and Instrinsic Value

The human being as described in late antique Neoplatonism—it matters little whether we speak of Plotinus or his opponents here—attains its goal of union with all things by communing with the divine intellect. Plotinus takes a rather different path than his opponents to the achievement of this goal, since he claims we have an interior tie to the divine mind through the undescended part of our soul. If we recognize this even in our day-to-day processes of reasoning, we may acknowledge that we are connected with all other human beings and with all organic wholes, including the cosmos itself. Even my sense perception of a tree is only possible because our souls have both descended from a single universal intellect. This unity is never a direct relation between two souls or bodies. I am quite distinct in soul and body from the tree that I see. The part of my soul that nourishes my body does not also nourish the tree, and my body is distinct from the tree in place. My unity with the tree is mediated by my contemplation of, and union with, the universal intellect. In more ritually minded Neoplatonists such as Iamblichus and Dionysius, my unity with other souls and bodies is typically mediated by the performance of rites (in Iamblichus), or in contemplation that occurs in the course of performed rites (in Dionysius). The microcosmic nature of the human in Maximus and Eriugena changes the character of this union with other souls and bodies. Because the human grounds both the sensuous and the intelligible, the created and the uncreated, it is able to see a thing as either sensuous or intelligible, created or uncreated. The human interpreter can and must run the gamut of the thing's possibilities, from its discrete sensuous nature to its union with the God who is beyond it. The human interpreter unites himself to the thing by reducing their difference to a point where the two are not separate, being united in the divine.

We find in Nicholas of Cusa the continuation of the microcosm doctrine of Maximus and Eriugena, but also the beginning of a changed view of human nature that has grave ramifications for our possibilities of union with other things. Cusa adds the developing science of one-point perspective to the medieval microcosm concept, and so can conclude that the world whose divisions are overcome in a human being is the world of *that* human being, different from the world of another animal or even another human being.[7] In other words, worlds are perspective-dependent. The analogy to one-point perspective opens the door to a geometrical conception of space. A multiplicity of beings, each with its own perspective, sees from a particular point in a space that is able to accommodate these different perspectives.

Nicholas of Cusa: One-Point Perspective

Cusa's early work, *On Learned Ignorance* (1440), presents a portrait of the human microcosm in fairly traditional terms. The topic of the microcosm comes up as he wonders what kind of creature is suitable for the incarnation of Christ.[8] Could Christ have become incarnate as a line? Cusa answers "no" on the basis of his understanding of the incarnation. The incarnation comes about when God—the absolute nature, containing all other natures—unites himself to a particular nature. Only a particular nature that has every level of reality in itself will be suitable for this incarnation. A line has no soul (it is not alive) and no intellect (it cannot think), and so if it were united to the absolute nature, it would become all inanimate natures, not all natures. It is a poor match, then, for the absolute nature. The best match will be "a median nature, which is the medium that connects the superior and the inferior." This is humanity: "it enfolds the intellectual and sensible nature and binds the universe together within itself, so that it is reasonably called by the ancients a microcosm or little world." Only human nature, which has both the sensible and the intelligible within itself, can accomplish the purpose of the incarnation: to "recapitulate" all natures within a single being.[9] Human nature has each level of reality within itself. The absolute nature has the ground of every creature in itself. The combination of the two in Christ allows him to take on every particular nature in taking on human nature.

Cusa's later work, *On Conjectures* (*ca* 1442), takes up some of the same themes as *On Learned Ignorance*, but without the theological context. The human microcosm reappears here, but its nature has subtly changed. In *On Learned Ignorance*, the microcosm chiefly allows human nature to participate in both the intelligible and the sensible worlds, while the fact that it "embraces all things within itself" goes more or less without comment or import. In *On Conjectures*, Cusa reverses the emphasis. Now participation in the different levels of reality is only a prelude to embracing "all things within itself." Every species is a unity, meaning it has something that distinguishes it from all other species. Human nature, for example, distinguishes itself as a unity by its linking of the intelligible and the sensible worlds. Cusa tells us that every unity is nothing other than a contracted infinity, where by contraction he means the compression of a more expansive nature within a less expansive nature. If God is the actuality of all things (an infinite nature), then a human being is that actuality compressed into the actuality of only one thing (a finite nature). Human nature is "an infinity humanly contracted."[10] A bear, likewise, is an infinity ursinely contracted. The human and the bear still have all things within them, but these things no longer have their own actuality. Instead, each thing exists only potentially in the bear. It is, Cusa would say, enfolded in the bear. The thing becomes actual only when the bear unfolds it from himself. The thing unfolded by the bear retains a relation to the bear, and so the bear experiences the thing only from his own perspective, and not in the thing's own actuality. The sum of the things unfolded by each creature is the world of that particular creature. The unfolding is the vocation of the creature, since the creature is a unity (meaning that it is a contracted infinity), and "the condition of a unity is to unfold beings from itself, since it is a being which enfolds beings by means of its simplicity." To be one is nothing other than to have a particular contraction of the whole, or perspective on the whole. Human beings, then, have no monopoly on unfolding the world from themselves, and so enfolding other beings within themselves. Every unity—whether the unity of a bear, a human, or an angel—has other beings within it. Human unfolding distinguishes itself from other forms of unfolding only in that it is capable of unfolding not just things within one level of reality, but everything: "the power of its unity encompasses the universe and encloses it within the bounds of its region, since it reaches out to touch all things by sense or reason or intellect."[11] Here we return to the familiar ground of *On Learned Ignorance*. Human beings have faculties associated with every level of reality: sense for bodies, reason for human souls, and intellect for the angels. The comprehensive character of our faculties allows us to unfold more beings from ourselves than any other species.

Because of the comprehensive character of human enfolding, we are in some way identical with God, the absolute actuality of all beings, and with the world, the aggregate of all contracted beings. This identity is only partial: "the human is God, but not absolutely, since it is human." The human and God both have all things in themselves, but God has them absolutely, meaning that he is their ground and perfection. Their contracted actuality becomes more actual as it approaches identity with God. The human does not have them absolutely, but it does have them in as much as they can be related to it, or be seen from its perspective. And so "the human is the world, but is not all things contractedly, since it is human. The human is, then, a microcosm or a kind of human world." The human being, like the world, has all things in itself, but the world has them in it in their own contracted actuality. The world is simply the aggregate of all things. The human being, on the other hand, enfolds all things in itself and unfolds all things *from* itself, giving them their own actuality so that they may be different from it.

Cusa specifically uses the language of place to describe the activity of humanity here: "the region of humanity encompasses God and the whole world by its human power." God and the whole world fall within the region of the human being, but only as related to the human. The difference between human and God is particularly manifest in the fact that human beings contract their universal being to the field of their own perspective. A human may be anything, but always in a human manner: "the human is able to be a human God and God humanly. It is able to be a human angel, a human beast, a human lion or bear or anything else." This raises the question of where my being actually is. Is it in the bear that projects me from itself? Is it in my own projection of myself? Is it in God? Cusa demurs from this difficulty here. He suggests that I am in the bear to the degree that my being can be contracted to part of the region of a bear. The bear is in me to the degree that it can be contracted to part of the region of a human being. I may be within the bear's region, and the bear may be in mine, but they are still two different regions. This difference of region prevents me from seeing anything from the perspective of the bear, since each being can only "unfold all things from itself within the circle of its region." My power of unfolding is limited to my region, just as the bear's is limited to its region.

We must distinguish the universe (*universa*) or world (*mundus*) from the region (*regio*). The universe is the totality of things, but we experience these things only from the region of our perspective. Our region is primarily the region of our species, but Cusa also speaks of the region of our nation and our rite. Our ability to unfold the entire universe, our prerogative as human beings, is thus restricted to, for example, a human, American, Christian mode of knowing. This restriction does not put Cusa together with Iamblichus and Dionysius, who say that we as human beings hold just one place in the hierarchy of beings, below the angels and above nonrational animals. Cusa follows the later tradition of Maximus and Eriugena, who establish human nature as prior to the hierarchic ranking of beings in the universe. We unfold the universe from ourselves, and so are prior to it. But Cusa combines the microcosm concept with perspective theory: the universe that we unfold is only the universe as seen from our perspective. And every being is able to do this. Human beings are special not because they unfold a world from themselves, but because their world has more things in it than the world of any other creature. A plant, on the other hand, can only unfold a world of inert or living non-sentient bodies, because these are the only levels of reality in which it participates. The chief feature of the microcosm concept—that it posits human nature as prior to the universe—is eviscerated in Cusa's development of it, since every creature is prior to the universe that it unfolds from itself. The remaining feature of the theory—that human nature contains every level of reality—simply specifies the field of objects in our universe, and explains why Christ became human and not some other creature. It does not shape our purpose as human beings or condition our interaction with the divine. This redirection of the microcosm has a great influence on several of Cusa's later works.

In 1453, the Turks sacked Constantinople, the capital of the Eastern Roman empire that had somehow escaped collapse for nearly a thousand years after the end of the Western empire. In its aftermath, Cusa completed two works, *On the Peace of Faith* and *On the Vision of God*. Both deal expressly with the nature and problems of perspective raised by Cusa in his earlier work, though only the former of the two appears to have been directly inspired by the sack. Cusa had long thought of the division of human beings into distinct regions and nations as an important stimulus to world harmony. In *On Learned Ignorance*, he notes three insurmountable obstacles to a perfect unity among human beings.[12] First, no individual of any species can possess all the perfections of its nature. If it did, it would immediately become God. This has happened once, in the case of Christ, who possesses all the perfections of human nature. All other human beings possess human perfections in different proportions: Solomon has more wisdom, Absalom has more beauty, and Samson has more strength. A second obstacle arises from our inability as human beings to form a final judgment on which of these perfections is the greatest. We find ourselves within a "diversity of religions, sects, and regions," and our standards of comparison differ according to this diversity. A Muslim living in North Africa will differ in her judgment of what constitutes human excellence from a Christian living in Italy. A third obstacle is that each of us knows only a few other human beings, so that we cannot even get a sense of the range of human perfection. The world is full of people with surprising characteristics unknown to us. Cusa raises these obstacles not to show the imperfect character of the sense world—as we might expect from Eriugena—but to show the providential care of God. God himself has brought about this diversity so that "each would be contented with himself and with his country—so that his birthplace alone would seem sweetest to him—and with the customs of his government, with his language, and so on, so that there would be unity and peace without envy." We learn to love what is near us by the strangeness of the distant. When we each attend to our own region, then strife between peoples ends and peace overtakes the world.

On the Peace of Faith takes a more complicated approach to the diversity of religions, sects, and other regions. The dialogue describes a vision granted to an anonymous man who, after hearing about the sack of Constantinople, prays to God with "many sighs" that the persecution caused by "the diverse rites of religions" be ended.[13] The man understands that the horrors rumored to have occurred during the sack have their cause in the very diversity of rites that Cusa had praised in his earlier work as leading to peace. In the man's vision, the angels assigned to the earth take their places in the assembly of the saints with God presiding over them. In telling us the role of these angels, Cusa modifies the traditional doctrine that each nation has an angel set over it as guardian, a doctrine we have seen in both Dionysius and Iamblichus. Cusa tells us that these powers "were set from the beginning over the individual provinces and sects of the world."[14] The angels are allotted specific nations because of their cultural or geographic identity, but also because of their religious identity: their sect. When the angels later bring back human representatives from their respective constituencies, and a dialogue between God and the saints ensues, we see an Arab speaking as a Muslim, an Indian speaking rather generically as a Hindu, a Greek speaking as a philosopher, and other nationalities speaking with the voice of their associated sects. God exercises care, through the angelic powers, for all the different sects.

The angels ask for this dialogue not so that all the nations of the world may convert to Christianity. Part of the problem with the present situation, as God himself puts it, is that men are "making others renounce their own religion, long observed."[15] Conversion here takes its place with killing as a harmful result of diversity in rites. God reminds his audience of the sense of unity and continuity that the possession of one's own rite brings. The angelic representative shortly explains why the value of a rite lies in continuity rather than truth, using terms that have now become quite familiar to us, though with a new political dimension we have not seen in the earlier Neoplatonists. Unity is always differentiated into multiplicity by means of a ranking. The single form of an oak tree, for example, is extended in space by a differentiation of its parts: roots below, leaves above. Among human beings this means a difference in class between the masses and the few who "have so much leisure that they can use their own free will to proceed to knowledge of themselves."[16] These few do not need a rite given to them at the level of sensible experience, since they are able to proceed through the knowledge that is within them. The rest, the larger portion of humanity, must labor to survive, and so do not have time to seek God. God takes care of such people by providing prophets, not the same prophets for all, but different prophets to different nations at different times. Cusa calls these prophets "kings and seers," meaning that they are both political and religious leaders. These prophets establish a form of worship and laws for the people, who can only believe what the prophets tell them, since they themselves have no leisure to seek the truth. They soon become subject to a characteristic of our earthly condition: that an "enduring custom, which is thought to have crossed over into nature, is defended as the truth." These are the people who fight for their faith as for the truth, not realizing that they are simply pitting one custom against another.

Cusa aims to solve the problem of religious strife without abandoning his earlier approval of diversity in customs by distinguishing religion (*religio*) from rite (*ritus*). The speaker for the angels recalls the positive result of diversity, and concludes that the many rites should not become one, though he is not as confident as Cusa in *On Learned Ignorance*. He says: "perhaps this difference between rites cannot or should not be removed, so that the diversity may promote devotion, when a particular region strives to be more vigilant in its ceremonies."[17] Diversity has the effect of an athletic competition—it encourages each rite to strive to be the best. But rite is not religion. The peoples of the world must be made to see that there is already "one religion in a variety of rites," so that they do not mistakenly attempt to convert practitioners of other rites to their own, believing that they have thereby established one religion. The dialogue between the representatives of the nations and the representatives of God has no other purpose than to make clear that there is already one religion, though we only see it from the perspective of our own rite.[18]

Where *On the Peace of Faith* contracts our ability to judge to the region of our own people and rite, *On the Vision of God* focuses rather on our inability to judge outside the perspective of our own species. In this regard, it revisits and extends Cusa's discussion of the human microcosm from *On Learned Ignorance* and *On Conjectures*. Cusa adds to the mix here a careful consideration of the way our contracted perspective differs from the absolute perspective of God. *On the Vision of God* takes the form of an extended contemplation of a metaphor. Cusa sent the treatise to the Benedictine monks of Tegernsee, and included with it a painting having on it the image of Christ. He advises the monks in his prologue to set the image up on a wall—he suggests the north wall. Then one monk should walk from west to east across the room, while another walks from east to west across the room. Each will see the eyes of the image turning to follow him across the room. Then the two monks must talk to each other, so that each believes, though he cannot see, that the image moved in both directions at once. *On the Vision of God* intends to flesh out the difference between what we see and what we must believe, for sight in this case turns out to be deceptive.

Cusa uses the icon and its viewers as a metaphor for the structure of reality and our perception of it. The icon, since it looks on everything within the room, is the image of God, who receives his name from the fact that he "looks at all things." Cusa follows Eriugena here in deriving the name "God" (*θεός*) from "to look" (*θεωρέω*).[19] The looking of God is both cause of the looking of each individual creature, and also nothing but God's being seen by the creature. The first of these claims depends on the difference in kind between the looking of God and the looking of each creature. The looking of creatures depends on "the passions of the organ and our soul," by which the same person may at one time look joyfully and at another angrily.[20] Absolute looking, however, "enfolds all particular modes of seeing in a similar and simultaneous way." Without this absolute looking, there can be no looking in one way or another. These latter modes are simply contractions of the absolute looking. The second claim, that the looking of God is nothing but God's being seen by the creature, is true also for the metaphor of the icon. The icon cannot be said to see except to the extent that it meets our eyes, and it is able to meet many eyes at once because it gives itself to be seen by each person in particular. Cusa claims that God, too, sees by giving himself to be seen: "what else, Lord, is your seeing, when you observe me with a pious eye, than your being seen by me?"[21] From Cusa's first claim we know that we cannot see without God. And "sight" here is simply a useful way of talking about existence. When we turn away from God, and so cease to see him, we cease to exist. Therefore, God does not see us when we turn away from him, since we no longer exist. His seeing, then, is nothing but his giving himself to be seen, which is nothing other than his giving us existence.[22]

A third conclusion follows from these first two. If God is absolute, then he cannot be other or different from anything, since otherness and difference "happen to the image which is from the exemplar," not to the exemplar itself. If God at the same time gives himself to be seen by contracted sight, then he must be seen under the same conditions as the creature that sees him, since otherwise he would appear different from that creature. "Every face, therefore, which can look at your face, sees nothing other or diverse from itself, because it sees its own truth."[23] Cusa is speaking literally of God as portrayed in the painting, in which he appears as a human being. But the painting falls short as a metaphor for God here because, if a lion were to come into the room and look at the icon, it too would see a human being, though it would see it from a lion's perspective. When we consider God in himself and not in his icon, we still give him human characteristics, since "a human being can only judge as a human." But the same goes, startlingly, for every animal species: "if a lion should attribute a face to you, it would only judge it as a lion's face. An ox would judge it as an ox's face. An eagle would judge it as an eagle's face."[24] Our approach to God then differs from our approach to creatures. The ox sees a tree in the manner of an ox, but it sees God *as* an ox. I see the tree in the manner of a human being, but I see God *as* a human being. Cusa comes to this conclusion after a consideration of the icon: "when I look at the drawn face from the east, it likewise appears that it so looks at me." My position relative to the icon conditions how it appears to me. If I want the truth, and not just appearance, I must get beyond my relative position. That is, I have to think beyond what I can see. The monks who, through mutual communication, learn that the icon does contradictory things at the same time arrive at a belief that they could never have acquired through their own sight. The human beings in *On the Peace of Faith* learn straight from God and his representatives a belief about religion that they could never have acquired from within their own rite, though presumably they could also have acquired it from mutual communication as the monks do in *On the Vision of God*. With this in mind, we begin to understand Cusa's motivation for studying the Qu'ran and other texts outside the compass of Christianity with such care. The unity of religion is not something Cusa can see for himself. It can only be believed, after communication with people of other nations and rites.

Cusa is the first author in our study to restrict our knowledge of all things as a unity to an acknowledgement that we do not know all things as a unity. We believe in it without seeing it for ourselves. For Iamblichus, Dionysius, and Maximus, the performance of rites is not simply an act of belief. The rites are the visible ground of a mythic context in which all things are gathered. Our performance of the rites gives us communion with God and the sharing of this context, however limited it may be by its restriction to a particular sacred place. This participation in unity is denied to the Cusan human being, which always sees from its own perspective. Our interaction with other beings lets us see the limitations of our perspective, but we cannot see beyond those limitations. The unity of different regions, like the universal religion, is a matter for belief and not sight.

Nicholas of Cusa: Homogenous Space

The perspectivism of *On the Peace of Faith* is a response to the brutality consequent on the contesting of the sacred place—manifested in the sack of Constantinople—but it has as its consequence the homogenizing of space, which carries with it its own perils. By conditioning our knowledge with the limitations of our nature and our nation, Cusa sets the stage for the replacement of place by space. If I see two things as belonging together, but I can imagine a lion (from *On the Vision of God*) or a Muslim (from *On the Peace of Faith*) seeing them as not belonging together, then I can imagine the two things as primarily and in themselves having position, while I or the lion or the Muslim contribute to them their place from outside. As a result, place may be conceived as subordinate to space, the neutral field in which objects are before they are incorporated into the region of a particular species or people. And Cusa does indeed tend to think of a homogenous space as the field within which beings take their positions.

Cusa's most famous assertions in *On Learned Ignorance* relate to the nature of the universe, though he derives them from traditional reflections on the nature of God. We have already seen such reflections in Dionysius' discussion of God as "greatness" and "smallness" in *On the Divine Names*. If we consider God as an extension, we must characterize him as both the maximum and minimum extension, for he will not be a relative but an absolute extension. An absolute is no longer a quantity, whether it be maximum or minimum, and so the two will coincide, since nothing distinguishes them outside the sphere of the relative. If God is the absolute maximum and minimum, then we must not be able to find any absolute maximum or minimum in any of the things within the universe that we measure with quantity. Motion is one of the things that we measure with quantity, so there can be no absolute minimum motion. This means that there can be no absolutely unmoved center of the universe, whether we put the earth at that center, or the sun, or anything else. Everything within the universe is in motion, even if we are not able to see it. If the universe has no center, it also has no circumference, since the circumference is defined relative to the center. Cusa himself does not employ this argument against there being a circumference to the universe, preferring instead to adopt the traditional Aristotelian argument against there being a place for the cosmos. If the universe had a circumference, then "it would have within itself its own beginning and end, and the world itself would be bounded by something else, and outside the world there would be something else and a place."[25] A circumference to the universe would mean that some body would exist outside the universe so as to serve as its limit. No body can exist outside the universe, since the universe is by definition all things. Cusa does think the universe has a place, but it is not a body. It is Plotinus' third kind of place: the first principle of all things.[26] Cusa's denial of a bodily circumference to the universe, then, does not break with tradition. It is his denial of a center to it that is radical.

Cusa's denial of a center to the universe has grave consequences for his theory of perspective, which he shortly brings into his discussion as a way to help us understand how the universe can have no center. He asks us to suppose that one person stands on the earth at the north pole, and another stands on the supposed circumference of the universe directly above the pole. To the person on earth, the person on the circumference will appear to be at the zenith of the universe, the highest point above his head. To the person on the circumference of the universe, the person on earth will appear to be on the zenith, again the highest point over his head. Cusa concludes: "wherever someone is, he believes himself to be at the center."[27] Cusa continues to apply his perspective theory in demonstrating that the universe has no natural places. The Neoplatonic tradition that precedes him understands the heavenly bodies to be of greater worth, and to occupy a greater region, than earthly bodies. To Iamblichus, the heavenly bodies are the visible bodies of the gods; to Dionysius, they at least have immortal characteristics, following a "wholly changeless path, without growth or diminution."[28] Cusa refutes five arguments in favor of the earth's imperfection, but for our purposes the most revealing is the fourth of these: Cusa's refutation of the argument "concerning place."[29] The argument claims that the inhabitants of the earth are less noble than the inhabitants of the stars, and so the earth itself must be less noble than the stars, and so we have reason to recognize a hierarchy of places within the universe. Cusa acknowledges that, if we treat the cosmos as an organic whole composed of parts, then there may be a proportion of regions in the universe just as there is a proportion of parts in the body. Just as some parts of the body are greater than others, so it is with the parts of the cosmos. But we cannot set up such a proportion between the parts of the universe, since we cannot see from the perspective of the whole. Nor can we compare ourselves to other inhabitants of the cosmos directly. To set up a proportion between inhabitants of different regions of the universe, we must have some knowledge of both species of inhabitant. We find, however, that our knowledge extends only as far as our own species. Even on earth, we and other animal species do not share a region and so know little of each other. Cusa says: "on this earth it happens that animals of one species, making as it were one specific region, unite themselves and mutually participate in what belongs to their region, because the specific region is common to them. About the others, they neither concern themselves nor do they apprehend anything truly." We humans share a region to the degree that we treat certain things the same way, but our region overlaps with the regions of other animals, which we do not share, and so know little about.[30] Cusa here assumes that animals other than humans have the capacity to communicate, and that we can understand them, but only to a limited extent. He says that "an animal of one species can apprehend a concept of another, which it expresses through audible signs, only in the case of a very few signs, and from outside. Even then, it takes long experience and is only an opinion." If our knowledge of other species in our own region has such limitation, we can know nothing at all about the inhabitants of other regions. We cannot say, then, that they are proportionate to us, and that we are less noble than they are.

From an analysis of the homogeneity of space, Cusa returns us repeatedly to the limitation of our perspective. The two theories are interconnected. My perspectival organization of the world around the point I occupy depends on there being a space capable of being occupied at every point. Likewise, the concept of a homogenous space means that every being sees every other being from a different relative position, a different perspective. Space now divides itself into sites for possible perspectives. Cusa is often wrongly credited with pioneering the concept of space as infinite and as homogeneous, though he denies that space is infinite, and is by no means the first to consider space as homogeneous.[31] Cusa does innovate in his application of the burgeoning science of perspective to the problem of space, and in this he is a true pioneer of the modern era. The combination of space and perspective leads already in Cusa to a confession of ignorance when it comes to adopting the perspective of other things. This ignorance has more recently posed an acute problem for our ability to value other things outside the use they have for us.

Intrinsic Value and Learned Ignorance

Take away the Hellenic world of sympathetic connections between people, things, and gods, and take away also the confidence remaining in the Christian Neoplatonists that we see things as they are. If we have no common context that invests things with meaning, and we see only the side of things that presents itself to our perspective, then the only comprehensible point of reference for the value we give to things is ourselves. The tree outside my window, for instance, is valuable to me because it provides shade in the summer and is a source of aesthetic pleasure year-round. These values are instrumental: they depend on the uses that I find in the tree. Because these values originate in comprehensible uses, I can explain them fully. If someone asks why I do not cut down the tree, I can tell them, so long as I refer its value always to myself. The term "value" sets us well outside the ancient and medieval philosophy we have been describing. In modern usage, the term almost invariably implies a relation to a perceiving subject. The subject either applies the value to a pre-existing object, or recognizes a value inherent in the object. In the absence of a perceiving subject, the object may have a nature and various characteristics but not value. While Cusa would not use the language of value, the perspectivism to which he contributes soon develops into a philosophy in which the language of value first becomes possible. A tree may be unfolded within the regions of a human and a bird, but these remain different regions, since the human and the bird value the tree differently. In the case of the tree outside my window, its berries are edible to the bird but not to me. This, at the very least, distinguishes the tree in the region of the bird from the tree in mine.

Though I can see the tree only from within my own region, I can acknowledge the value of the tree for the other species that have it as part of their region. I can see that the birds and squirrels both rely on the tree's berries as food in the wintertime, and so acknowledge that it possesses instrumental value for the birds and squirrels as well as for me. I am still the one who establishes the value of the tree, but I establish it with reference to the birds and squirrels rather than myself. In this case, the valuer differs from the one for whom the thing is valuable. The one for whom a thing is valuable could be a human being, an animal, even a plant—anything for whom something else can be useful. Such a one does not have to be able to recognize and explain the value. This latter task belongs only to the valuer, the human being who identifies the value of the tree for the birds and squirrels. Despite my recognition of the tree's value for birds and squirrels, however, I may still cut it down.[32] Something more is needed if what is valuable for a bird is to become valuable for me.

This problem is easily solved if the bird itself has instrumental value for me, for example, as a source of aesthetic pleasure. In this case, the tree will have indirect instrumental value for me as a source of nourishment for the bird who is my direct object of interest. But there are other values I may give to a thing beyond the directly or indirectly instrumental. The thing may have a place in the mythic context that gives identity to my historical people. The laurel trees growing within the walls of Apollo's oracle at Didyma may have little instrumental value for the worshippers there, but no one would ever think of cutting them down. The trees are valuable not for some service they provide to an independent valuer; they are part of the very identity of that valuer. To cut such a tree down would be to lose a link to the context in which the Greek and his actions have meaning. And so those Christians of Late Antiquity intent on destroying the Hellenic world gave special attention to the cutting down of sacred trees. Things may also possess value because of their place in contexts smaller than the all-encompassing mythic context that identifies a historical people. A thing may be valuable to me because it identifies me as a member of a group or as a distinct person. If the tree outside my window is an important part of my life now, and I move away, and return twenty years later to find that the tree has been cut down, I may feel a painful sense of absence as though a part of myself has disappeared.

The value possessed by a thing as the ground of a mythic context is easy to recognize but more difficult to explain, perhaps because we can draw no sharp distinction between the thing that is valued and the person who values it. How can I explain that the tree is a part of me, since we are two discrete beings, occupying different positions in space? We overlap only in a context that is not directly visible. Perhaps because it is so difficult to explain, this "mythic value" is often reduced to a form of instrumental value. The worshipper of Apollo at Didyma may say that the reason why the laurel trees there should not be cut down is not because they are part of the identity of the Hellenic people, but because Apollo will become angry and kill anyone who cuts down one of his laurel trees. A clear instrumental value thus attaches itself to the preservation of the trees in place of the less articulable value of identification.

If the tree has no comprehensible value to me at all, neither direct nor indirect instrumental value, nor the value of identifying me as a person or member of a group, I may still insist that the tree has an intrinsic value and ought to be preserved on that basis. Some such value motivates Christopher Stone's famous 1972 article "Should Trees Have Standing?—Toward Legal Rights for Natural Objects," subsequently published as a book.[33] The concept of intrinsic value has become a contentious issue in environmental ethics, largely because it is not always distinguished from the forms of value we have already identified. Even when it is carefully distinguished from instrumental value, intrinsic value remains a problem because of the difficulty in explaining it. We cannot answer the question of why the thing is valuable, since a "why" question presupposes a ground of the value external to the thing itself. We may then simply abandon the project of attempting to explain the intrinsic value of a thing, and leave the value as a matter of recognition and assertion rather than articulation and defense. Or we may attempt to couch a thing's intrinsic value within a larger system of value, without reducing the thing's intrinsic value to an instrumental value for the system.[34] Organs in the system of the human body, for instance, have no intrinsic value. The heart and liver are intrumentally valuable as parts of the larger whole, and have no value without reference to that whole. What about a city? The human beings who make up the city play roles that are necessary to its proper functioning, but most of us would agree that each human being also has an intrinsic value irrespective of what he or she does for the city. But the city does not ground this intrinsic value. It simply provides a context in which its citizens can take on determinate identities. One person may become a farmer, and find himself in the context of farming. The city, in this case, establishes what we have been calling "mythic value," and has no bearing on intrinsic value. The same goes for the ecosystem, the natural analogue to the city. To the degree that we find value in a plant by identifying it as benefiting from and contributing to other plants, animals, soil types, and climate, we do not address the plant's intrinsic value at all. We simply establish a context in which the plant may share meaning with other things.

What must we do then to find a context in which intrinsic value may become articulable without destroying the very value we wish to articulate? Cusa provides us with a clue. We have described someone with intrinsic value as the sole reference point for value in things, as the unfolder of a region. I recognize myself as the sole comprehensible point of reference for the value I give to things in my region. That is, the region unfolds from my perspective. But I may come to believe that my perspective is not the only perspective. I may, for instance, see that the bird values the tree instrumentally in the same way that I do. I can never acquire the perspective of the bird, but I can come to believe that it has its own perspective by observation of it. I use the term "perspective" here in a broad sense not limited to sentient beings. The tree, too, has its own perspective, because it composes a region of things valued with reference to itself. As the sole point of reference for a region of values, the tree itself has intrinsic value, just as do the bird and I. I can recognize this intrinsic value in the bird and tree only if I believe them to have their own regions of value, and I can believe this only if I allow that my region is not the only region. A twofold learned ignorance is at work here. I must acknowledge my ignorance of the bird's perspective, but I must also acknowledge my ignorance of the thing in itself which finds its way into both our regions: the tree, for instance, which nourishes the bird but would poison me. And yet, I must acknowledge that there is such a thing in itself if I am to account for the simultaneous presence of the thing in multiple regions at once. In other words, I assume the thing as object of an absolute perspective, even as I myself perceive it only from my own perspective. Such an absolute or divine perspective must remain absent and incomprehensible to me, for it would enter into contradiction if it became present and comprehensible like a city or a human body. The same tree would, in itself, be both poisonous and nourishing. So the divine perspective does not take up the relative perspectives into a system, but stands absently and incomprehensibly behind them, and so is able to ground intrinsic values.

Can we attribute intrinsic value to places as well as things? One place in particular has frequently been accorded intrinsic value: the wilderness area. Michael P. Nelson includes an intrinsic value argument as one of his thirty collected arguments for wilderness preservation in the 1998 anthology, *The Great New Wilderness Debate*.[35] Incorporating the work of E.O. Wilson, Edward Abbey, Holmes Rolston III, and William Godfrey-Smith, he follows the approach to intrinsic value that abandons the attempt to explain it, claiming, "'wilderness' defense needs no articulation. Designated wilderness areas just *are* valuable." He notes that "such locales, then, join the list of other things, like friends, relatives, children, family heirlooms, and so forth, whose worth is not contingent upon anything other than their mere existence, whose value is intrinsic."[36] All of these examples save that of the family heirloom refer to human beings, whose intrinsic value I can recognize by analogy with my own, and so are not good examples of things that "just are valuable." The family heirloom example fares no better, since the family heirloom does not possess intrinsic value. Take the example of a pocket watch, which has been handed down over several generations, from father to son, within a family. As a pocket watch, it possesses instrumental value; it allows me to know what the time is. It becomes a family heirloom by its long association with a loved one, and beyond that with barely known ancestors who nevertheless have an intimate connection with me. I value the watch because it is a visible link to my identity with these people. If instrinsic value means to be valuable without reference to anything external, then this watch does not possess intrinsic value. It possesses the "mythic value" of things, which visibly ground our identity with a group, in this case, my extended family, past and present.

Places, too, can acquire such mythic value. Mircea Eliade points to this acquisition of value on the part of a place as a "degradation and desacralization of religious values and forms of behavior,"[37] because such a place becomes significant to a person without making an absolute claim on him, as would the sacred place that grounds the context of an entire historical people. The place that identifies a person is not now the location of a theophany, but of a momentous private event in his own life. Examples include "a man's birthplace, or the scenes of his first love, or certain places in the first foreign city he visited in youth." The value of such places depends entirely on the person who invests them with value, and each place retains such value for a relatively small group of people: a family, or a group of friends. Such a value is typically not strong enough to preserve the place from transformation or destruction. Because the meaning of the place is relative to a small group, and because the group understands that the place is meaningful only to it, Eliade is right to see this valuation of place as a degradation of the universal and absolute significance Iamblichus finds within sacred landscapes and Dionysius finds within the walls of the church.

A place, then, can have mythic value. If it is to have intrinsic value as well, one that I can explain to myself and others, then we must have some basis for identifying it as the site of a perspective. We have seen Cusa explain that only a unity can unfold a world from itself, and so have a perspective, since a unity is nothing but a contraction of the whole. It thus has all things within itself, but it unfolds them in accordance with its own nature and circumstances.[38] A pile of rocks does not seem to possess such a unity, since the particular rocks do not gain their identity from being part of the pile. Each particular rock could also be split into smaller rocks, and these smaller rocks would lose nothing of their identity from no longer being part of the larger rock. The same goes for the yard outside my window, and a bend in a river. A chunk of turf, or a cup of water, lose nothing if they are relocated from the yard or the river. My arm, on the other hand, loses its function altogether if it is taken by itself and not as a part of my whole body. It is no longer an arm, properly speaking, when it has no relation to the rest of the body. As the whole of which the arm is an inseparable part, I have the unity necessary to merit intrinsic value. Places, on the other hand, do not. The value of a place is instrumental to the things that are in that place, or else its value arises from the share it has in our lives. We value the place, then, only because we value the things that are in it, or because it has somehow become a part of us. In the former case, what we are calling a place would be more adequately characterized as space. The yard outside my window provides a neutral field in which the bird and tree may unfold the worlds of their perspective, and I may unfold mine.

Hiddenness and Learned Ignorance

Though places may not have an articulable intrinsic value, we find in Heidegger's middle and later work a parallel concept in the "hidden" (*Geheimnis*). If we recognize the hiddenness of places, then this hiddenness, like intrinsic value, may prevent us from reducing them to their usefulness for intrinsically valuable human beings.

Heidegger begins to explore the concept of the hidden in "The Origin of the Work of Art," where he calls it the "earth."[39] Whenever we produce a tool for some use, we always employ some material. Heidegger gives the example of an ax. If I am going to produce an ax, I need stone. But I do not need the stone *as* stone. I need sharpness and durability. The ax becomes more efficient to the degree that the stone "disappears into usefulness," by becoming completely absorbed into the utilitarian context of the ax.[40] When a farmer needs to chop wood, he reaches for the ax, not stone. He may not even see the stone except as part of the ax. The work of art, on the other hand, does not cause the materiality of its material to disappear. "The sculptor uses stone just as the mason uses it, in his own way. But he does not use it up."[41] The stone appears in the work of art *as* stone. Even if the sculpture is of Athena, and so is the visible ground of the mythic context of the Greeks, the stone is not completely absorbed into that context. The work of art not only sets up a world, by grounding a mythic context, but also "sets forth the earth," by allowing the material to show itself as material.[42] Because the earth is, by definition, other than the world, whether that world be mythic or merely utilitarian, the earth cannot be explained. The ax can be explained in terms of its purpose, but the stone cannot. The earth set forth in the work of art is "essentially self-secluding." If we attempt to give it a purpose or simply make it comprehensible, we destroy its character as earth. As an example of the latter, Heidegger provides the quantitative analysis of modern experimental science. To make the heaviness of the stone comprehensible to us, we quantify it, by "placing the stone on a balance." Now "this perhaps very precise determination of the stone remains a number, but the weight's burden has escaped us." A number gives us the illusion of having understood the heaviness of the stone, when in fact it has only made the stone more manipulable, more likely to be absorbed into an ulterior context, and so more likely to disappear *as* stone. Measurement, in other words, has a dematerializing effect. Measurement makes things explicable, and the material is what cannot be explained. It must show itself as "self-secluding" if it is to remain material.

In his Parmenides course, Heidegger addresses the nature of places in terms of this opposition between a self-secluding earth and the humans who seek to comprehend and manipulate it. He does this briefly, in passing, using an example of a place that had preoccupied him for an entire lecture course the preceding summer: the river.[43] He has just described the "aggressive making secure in advance" which characterizes the rational method, a method whose origin he has earlier found in Plato. Here he notes the visible changes effected to the river by the rational method: "a river no longer flows in the hidden (*geheimnisvollen*) course of its windings and turnings along banks it itself has carved out, but it now only pushes its water to an 'end' predirected to it without detours, between the uniform rails of cement walls, which are in no way banks."[44] The river hides itself first of all by its "windings and turnings," which prevent the sailor from seeing around the next bend. But the river also hides itself by determining its own course, following "banks it itself has carved out." The sailor may navigate it in a boat, and the farmer may draw water from it for irrigation, but neither changes the river's self-determined course. They use the river for their own purposes without reducing it to those purposes, with the result that its hiddenness remains apparent. Straightening the river, by building concrete walls between which it must now flow, eliminates both forms of hiddenness in favor of greater manipulability of its water. The river now conforms to a geometrical model—the line—that prevents it from determining its own course, and makes it easily accessible to the eye. The straightening of the river also facilitates the general measuring out of the landscape on a geometrical model: farmland may now be arranged around the river in rectangular plots.

In his 1950 essay "The Question Concerning Technology," Heidegger revisits the river in the context of human building, in this case the building of bridges and dams. A dam on the Rhine river reduces that river to a purpose, since it "is not built into the Rhine River as was the old wooden bridge that joined bank with bank for hundreds of years. Rather, the river is dammed up into the power plant."[45] The bridge opens up a context in which human purposes are accomplished: it allows human beings to gather by crossing the river. The farmer on one side may gather with the villagers on the other side, and in the context of that gathering vegetables may be sold, news exchanged, and so on. But this gathering does not eliminate the hiddenness of the river, since the bridge gathers the river, too, as Heidegger says in the "Building Dwelling Thinking" essay of the same period. Or, as

Heidegger puts it here, the bridge is "built into the river." It allows the river to be itself, setting its own winding course. When it is dammed, the river *as* river disappears into the purpose given to it by the dam: "what the river is now, namely, a water-power supplier, derives from the essence of the power station." What disappears from the river here is more than its material, as Heidegger suggested in "The Origin of the Work of Art." The river has lost its ability to appear as more than an instrument in the utilitarian context of human beings. We see this change directly in the appearance of the river. If its water is now a resource for irrigation or drinking water, it is straightened or even disappears into a pipeline in the interest of getting its water to its consumer as quickly as possible. If the river is now a power source, it disappears into a reservoir of water. In either case, the river loses its ability to rest on its own ground.

"Ground" is Heidegger's equivalent in his 1957 lecture course, "The Principle of Reason," for what he earlier calls the "earth," though he equates the two even in "The Origin of the Work of Art," saying that the earth *is* ground.[46] The ax fails to set forth the earth because it does not present the stone *as* stone. The stone does not rest on its own ground, but is the servant of the purpose given to it by the ax. The context of the 1957 lecture course is somewhat different. Heidegger is concerned not with the subordination of things to human purposes, but with the presumption that things can be understood in themselves in terms of purposes, human or otherwise. This presumption necessarily precedes the subordination of things to purposes that concerns him in his earlier works. Here Heidegger takes as his example not stone or a river, but the rose discussed by Angelus Silesius in his poem. Heidegger raises the possibility that what Silesius means when he says that a rose lives without a why is that things like roses "live because of reasons and causes, but never *according to* reasons." The rose does not come up with reasons for its blooming in advance and then act according to those reasons. The rose is not conscious of itself, and so it can think of no "why" for its blooming. There are still reasons for its blooming, and we as human beings can find out those reasons by studying botany and ecology. But if Silesius "only wanted to bring the difference between the rose and humans into relief, then he could have said: 'the rose blooms because the sun shines and because a lot of other things surround and determine it.' "[47] Silesius has not said this. He says: "the rose blooms without a why." Silesius has seen that to explain the rose's blooming fully in terms of causes and purposes, even if they are not human purposes but serve the rose itself, is to prevent the rose from resting on its own ground. It does not reveal the rose *as* rose, but only in its relation to other things. If we wish to see the rose as a rose, we have to refrain from explaining it, since explanations always explain away the very nature of the thing concerned. We must adopt a form of learned ignorance, a recognition that things can present themselves to us *as* things only when we allow them to retain their essential hiddenness.

Heidegger is able to include more in the category of the hidden than we were able to include in the category of the intrinsically valuable because, unlike the intrinsically valuable, the hidden need have no special unity in itself. Intrinsic value, if it is to be articulable, can only be attributed to things that are capable of a perspective, meaning that they can unfold a world for themselves. A river's unity is too adventitious to make it a candidate for intrinsic value, but it may still be hidden, since its hiddenness does not depend on its possession of an interior unity. Human beings gather the earth into a place like the bridge, and then it may appear as a river. Prior to the appearance of the river in a place, we can consider the river only abstractly and outside its true nature—on a map, for instance. In itself, if we may speak this way, it is simply the earth. A rose, too, is earth until that earth is gathered into a rose, when we let the rose be a rose. Heidegger's human beings have as their distinctive vocation the gathering of earth and sky, mortals and immortals into things and places, and both things and places set forth the hiddenness of the earth.

Heidegger does not go into detail on how the earth is gathered—indeed, he cannot go into great detail, since he treats the earth both as the material of bodies and as the incommensurable material of language—but we may go into greater detail if we restrict our discussion to the gathering of earth into bodies.[48] The bridge that gathers the river and its banks in a landscape has the character it does because of who we are: beings of a certain height, a certain pace, a certain capacity of sight. If our stride were more than two hundred feet, for instance, we would need few bridges at all. These bodily determinations help to shape the character of the bridge. But the bridge is also shaped by the meaning we give to it, a meaning determined not directly by the body, but by the context in which we use the bridge. At the very least, the bridge allows human beings a crossing from bank to bank. If this crossing is made solely to get produce from field to market so that it may be sold, the material of the bridge disappears into its usefulness. But if the bridge is allowed to be part of the identity of the human beings who use it, then it begins to gather the earth in gathering these human beings together. The stone is no longer used up in the bridge, but comes to appear as stone. If the context of the bridge is a mythic context, then it will allow the gods to appear as well, through the act of consecrating the bridge to a god or to a saint. We would not gather the earth into the bridge unless we found ourselves in this context, but the context itself is shaped by the way our bodies gather the earth. Even the gods take shape in and through the gathered earth, and so appear relative to our bodies.

When our language is no longer capable of naming the gods, as in Heidegger's later essays, the gathering of the earth does not allow the gods to appear in their own form, but in the hiddenness which that gathering makes apparent. In these essays Heidegger stands much closer to Meister Eckhart, whose God is a "quiet desert," than to any of the early medieval figures of this study, save perhaps Eriugena. In the absence of a world of myth, Heidegger's sacred places do not provide the communion with all things and the divine that the early medieval Neoplatonists held as the essence of the sacred place. Instead, they make apparent the ineradicable distance of gods from mortals, by revealing the gods' absence from meaning. Yet the renewal of the world of myth that would allow the gods to return to Heidegger's sacred places raises all the specters of the historical people that we see in Heidegger's Germany and Eliade's Romania. As it happens, a new mode of union has appeared in the absence of place and the cosmically connected human being that dwells in it.

The Affective as Consolation for a Vanished Place

So long as we treat human nature as having positive access to the universal, we may still find communion with other things and with the divine. The stillness of the sacred grove is able to compose the human mind because the mind and the grove share a common stillness. The bread and wine are able to unite the faithful gathered in the church because they visibly ground the common mythic context of the faithful. When we break the final tie with things in the world by removing the universal intellect that unites us to them, these connections disappear. This is not to say that we can no longer enjoy the stillness of a grove, but we are precisely *enjoying* the stillness, rather than uniting ourselves with the universal intellect in the grove. If this enjoyment gives our minds a certain composure, all the better, but the enjoyment and not the cosmic intellect is now the cause of this composure. Likewise, we may still get a feeling of unity from the eucharist, but again, we are now speaking of a feeling. Emotion now serves as the mediator between other beings and ourselves. This change of mediation from intellect to emotion has colored the twentieth-century rediscovery of place.

Iamblichus, Dionysius, Heidegger, Gadamer

Porphyry is well known to have suffered from extreme emotions. He himself tells the story of how he was considering suicide when his teacher Plotinus came to him and, in Porphyry's words, "said that this desire did not spring from an intellectual disposition, but from a disease caused by black bile, and he commanded me to do some travelling."[49] Porphyry is well aware, then, of the power that dispositions of the body exercise over our experiences. In his letter to Iamblichus, he suggests that such dispositions are responsible for the unusual experiences of the priests when they undertake divination. Iamblichus responds by sharply distinguishing divinatory enthusiasm from a passive experience (πάθος), stressing that enthusiasm is a change of being: "neither is enthusiasm simply self-effacement (ἔκστασις), for it is a reelevation and transition to a more excellent condition of being." Mere self-effacement need not mean the elevation of one's being. It could simply mean going out of one's mind at the prompting of external stimuli or changes in body chemistry. Enthusiasm does involve external factors, like the sights and sounds of the rites, and physiological changes brought about by fasting and ritual actions, but these do not cause the enthusiasm. If the proportion of humors in the body or any other change in its physical constitution is the author of the enthusiastic ecstasy, "it will be a bodily passion of perversion, arising from natural motions,"[50] and natural motion cannot accomplish a supernatural end.

Porphyry also points out the more specific example of music, used by many religious traditions to induce an altered emotional state. The Corybants, as well as the devotees of Sabazius and Cybele, all use music in their rites, apparently with the goal of inducing the self-effacement necessary for religious union. Iamblichus acknowledges the great power of music. He himself, in his biography of Pythagoras, says that Pythagoras one night prevented a youth from killing his mistress in a state of passion by convincing a piper in the vicinity to change his mode from the Phrygian to the Spondaic. Once the piper had done this, the boy's soul "was rearranged without delay, and he went home in good order."[51] The Pythagoreans themselves "purified their reasoning from the disturbances and uproars of the day with certain odes and kinds of melody, and by this means they prepared for themselves a still sleep with few dreams and good ones at that," and on waking up the next morning they rid themselves of drowsiness with chants of a different kind. These effects of music, Iamblichus says, are "natural, human, and the works of our own craft, and nothing at all divine is manifested in them."[52] The natural power of music, as he has just described it, can inspire a feeling in us, and so may be the cause of action in us, but it does not change our degree of being. It is not divine, and so has no direct relevance to the Hellenic rites. We use music in rites not to inspire feeling, but because "there is suitably given to them a nature shared (συγγένεια) with each god's own rank and power, and with the motions in the totality itself, and with the harmonious sounds that whistle from these motions."[53] Certain forms of music, like certain things, have a symbolic connection with the gods, and this connection unites them to the gods' own nature. The gods hand down this communion through the highest form of music—the music made by the heavenly bodies as they rotate in the cosmos—all the way down to the music played in the rites. The ritual power of music comes from this communion, not from its natural power to move the soul. If the soul seems especially moved by the natural power manifested in ritual music, it is only because "the soul, even before it gave itself to the body, listened to the divine harmony. And so, when it arrived in the body, and heard such melodies as best preserve the divine trace of harmony, it embraced these." The soul may revel in the trace of the divine present at the level of the visible in the audible harmonies played during the rites. This harmony is not symbolic, but comes from the fact that harmony here is a natural image of the higher harmony of intellect. Iamblichus guards against a view of ritual music as cathartic, drawing out the impure emotions of the soul so as to leave it purified. If music were a "purgative, purification, or medicine," then its effect would depend on a "disease, excess, or superfluity" in the soul. A healthy soul would then receive no benefit from music. As it is, "its every principle and foundation are established from above, as divine."

Dionysius has no Porphyry to question him about the emotional power of the rites, and so he never denies such a power. That he would deny it, if asked, seems clear from the fact that he never mentions the affective side of music, fragrance, or any other sensuous characteristic of the rites. The singing of psalms and other hymns unites the community of the faithful in the sacred place, but it does not need to produce a feeling of unity. What matters is the change of being. The things to be contemplated during the service—the fragrance of the incense, the sight of the bread and wine—serve as symbols to draw the mind to the intelligible, but Dionysius never describes this drawing as affective, dependent on a feeling produced by the fragrance or by the beauty of the bread and wine. The rites have their power as an image of Christ's historical incarnation and his eternal divine character. The affective is conspicuous only by its absence from Dionysius' interpretation of the ecclesiastical rites.

In his rehabilitation of place, Heidegger carefully removes its importance from the affective. The establishing and sustaining of architecture, and art in general, has being rather than feeling in mind. We have seen how Heidegger in his Parmenides course carefully differentiates the Greek experience of the extraordinary from our own experience. He says: "for those who came later and for us, to whom the primordial Greek experience of Being is denied, the extraordinary has to be the exception, in principle explainable, to the ordinary; we put the extraordinary next to the ordinary, but, to be sure, only as the unusual (*Nichtgewöhnliche*)."[54] For the Greeks, on the other hand, the "ordinary itself, and only insofar as it is the ordinary, is the extraordinary." Because the Greeks do not divide the ordinary from the extraordinary as two different realms of experience, the extraordinary does not present itself to them in exceptional ways, as "the monstrous or alarming,"[55] or "impressive and moving."[56] The daimons, carriers of the extraordinary, do cause the dispositions of horror, fright, joy, mourning, terror, as well as every other "essential affective disposition,"[57] but Heidegger warns us that "these 'affective dispositions' are not to be understood in the modern subjective sense as 'psychic states' but are to be thought more originarily as the attunements to which the silent voice of the word attunes the essence of man in its relation to being." We must not interpret the Greek descriptions of joy, mourning, and terror in the face of the daimonic to mean feelings as we now understand them. They are attunements of the silent word, which dispose the Greeks to see and understand the world around them in a particular way.

A raft of recent thinkers in the Heideggerian tradition has taken up his position in the broader context of interpretation. Lindsay Jones, who takes the work of Hans-Georg Gadamer as his point of departure, suggests that Gadamer would follow Heidegger in locating the power of architecture in change of being rather than "simply shifts in attitude or modifications of one's state of mind.. . .For Gadamer, experiencing art and architecture facilitates actual 'transformations in being.'"[58] Gadamer himself does not speak of change in being in the context of architecture, and when he does speak of art in general as causing a change in being, it is the thing represented by the art and not the viewer of the art that changes. Gadamer suggests that the work of art "is determined ontologically as an emanation of the original. It is of the nature of an emanation that what emanates is an overflow. That from which it proceeds does not thereby become less. The development of this idea by Neoplatonic philosophy. . .is the basis of the positive ontological level of the picture."[59] The Neoplatonists speak of causality as the emanation of images by their archetypes. The archetype does not lose any of its own reality in the production of its image, just as a woman who looks in a mirror retains all of her reality even as she reflects an image of herself. Gadamer concludes in the case of art that the archetype actually increases its being by the production of images. When the original is represented through the work of art, we find that, "through the representation, it experiences, as it were, an *increase* in being."[60] Jones does not discuss this treatment of art in Gadamer. He prefers to work from Gadamer's more general theory that things find their being only in interpretive contexts, and so concludes in the case of architecture: "buildings have no autonomous, stable being (or existence) outside of their interactions with human users." This may be true, but it is not Gadamer's emphasis when he speaks of art. Gadamer explores the character of the work of art in itself, while Jones prefers to focus on the effect of art on the person who encounters it. Jones allows "change of being" as one possible effect of architecture, but he notes two others: a "transintellectual (or at least transconscious) transformation," and "religious transformation or *spiritual awakening*."[61] As Jones describes them, these latter two relations to architecture bring in the modern aesthetic and psychological tradition, with its emphasis on feeling rather than being, and take us into a new context.

Disinterest

While Heidegger and several other recent thinkers on the sacred place have, in an apparently unintentional homage to the Neoplatonists, sought to remove it from the realm of feeling, the modern aesthetic tradition with its emphasis on feeling lives on in others. Pierre Hadot merits special attention in this respect, as he has undertaken a serious and lifelong study of the ancient world, and at the same time has applied to it a self-consciously modern aesthetic. We have already relied on his discussion of the landscape tradition in antiquity. His distinction of antique landscapes into charming and sublime has guided our exploration of landscape in Hellenic and Christian Neoplatonism. We do not find an explicit distinction of this sort in the ancients apparently until John Chrysostom, who divides objects of wonder into those that provoke no fear, like "the beauty of columns, or the works of a painter, or bodies in their bloom," and those that do provoke fear, like the ocean's depth. But Hadot gets his distinction of charming and sublime from the post-Enlightenment philosopher Immanuel Kant (1724–1804) rather than Chrysostom.[62] Hadot's reliance on Kant's form of the dichotomy subtly superimposes the modern strife between use value and aesthetic value onto the more ancient dichotomy. In other words, Hadot's distinction of landscapes into charming and sublime depends on a prior distinction, also Kantian, between interested judgments (of use value) and disinterested judgments (of aesthetic value).[63] Hadot couches this more significant distinction in thoroughly ancient terms. The first term of the distinction, he says, is symbolized by the god Prometheus, who "represents the craft that steals the secrets of nature from the gods who hide them from mortals, the violence that seeks to conquer nature so as to improve the life of men."[64] Though Hadot cites ancient sources for this symbolism, the Kantian theme of interested judgment shortly surfaces. Hadot tells us that "mechanical craft is never disinterested," and that the Promethean "relation to nature is never disinterested." The second term of the distinction is symbolized by the rural life. As described in much ancient poetry and philosophy, the rural life involves knowing the rhythms of nature, and living without a great deal of artifice. So it involves both a contemplative and an active side. Its essence is "to contemplate nature as it is, to describe it in language, but also to live 'according to nature.' " The Kantian theme of disinterested judgment shortly arises as the characteristic of our "contemplation of nature as it is." The attitude toward nature adopted by the poet of the rural life "implies also a disinterested regard directed at nature, a regard which can, for example, define a landscape." Here is the source of the charming landscape in antiquity: the disinterested or aesthetic contemplation of nature.

To nominate disinterested contemplation of nature as a goal for human beings is a thoroughly modern resolution of a thoroughly modern problem. The problem arises from the increasing organization of the world around us for utilitarian ends, and from our seeming inability to relate to the world in any other way. The aesthetic solution simply eliminates our use of the world and our interest in the world in favor of a disinterested contemplation of it. This sets the stage for the development of the modern aesthetics that Heidegger and Gadamer disown.[65] The aesthetic attitude does not free us from a world reduced to its use value, since the aesthetic is simply the inversion of the interested. I can look at the rose as a source of perfume, or I can eliminate my contemplation of its use and simply let it confront me with its form, but in neither case do I get beyond the polarity of use and its absence to a positive interaction with the rose's own nature. I do not experience the world in its own right, but only as the absence of objects in their usefulness. As we have seen, the Neoplatonists of Late Antiquity are not faced with this problem because they have a means of union with the nature of the thing that has disappeared by the time of Cusa.

Although Hadot is not bashful about bringing Kantian distinctions into his discussion of the ancient landscape tradition, he does wish to set that tradition within the parameters of its own philosophical distinctions. To this end, he frequently refers to a passage from the Stoic philosopher Seneca, who claims to "spend a great deal of time in the contemplation of wisdom. I look at it with the same stupefaction with which, on other occasions, I look at the world; this world that I quite often feel as though I were seeing for the first time."[66] Sometimes Seneca sees the world in the same way he sees wisdom, as a *spectator novus*, someone seeing for the first time. When he looks at it this way, it stupefies him. Hadot interprets Seneca here to mean that the "habitual, utilitarian perception" of the world drops away and we are able to look at the world without desire for all the things that it can do for us. If we interpret "habitual, utilitarian perception" to mean planning to use the world so as to maximize our own pleasure, then we see why the elimination of this perception belongs in Stoic philosophy. The more ascetic Stoics advocate the elimination of desire, which seeks the pleasurable, because desire clouds our ability to see and do what we ought to do.[67] The difficulty in identifying the Stoic dispassion with Kantian disinterest is that the Stoic's dispassionate contemplation of the world is simply a prelude to action in the world. The elimination of desire simply reveals to the Stoic the duties he has, and a new motivation for acting to fulfill them: natural impulse. Kantian disinterest, to the degree that it remains a contemplation of nature, does not reveal a motivation for action. It stands opposed to action and the interest that seems inevitably to characterize it.

Hadot attempts to get beyond this dichotomy of active interest and contemplative disinterest by supplementing his account of aesthetic perception with an account of artistic creation. The act of producing a work of art brings with it a feeling of unity with the forces that produce works of nature. He acknowledges first of all that the mathematization of nature has prevented us from representing it to ourselves as a world or cosmos: "the quantitative universe of modern science is totally unrepresentable, and within it the individual feels isolated and lost."[68] Like the number line, the modern universe goes on forever, and so our minds cannot conceive it as a whole with its accompanying circumference and center. We cannot locate ourselves within it, and we cannot think of it as a power capable of knitting all things together into one world. Nature is now a passive background to human action. She "is nothing more for us than man's 'environment'; she has become a purely human problem, a problem of industrial hygiene."[69] The reduction of nature to "environment" has not, however, eliminated the possibility of union with other things within the universe. It is simply that this union must occur by other means. Hadot poses the rhetorical question: "does not modern man, too, have his own experience of the world *qua* world?" He answers with a "yes," and the means to this experience is now an aesthetic perception not conceived as divorced from the perspective of the cosmos. He explains: "a disinterested, aesthetic perception of the world can allow us to imagine what a cosmic consciousness might signify for modern man."[70] Hadot introduces the cosmic consciousness here as a way to get around the apparent worldlessness of disinterest. Where a disinterested judgment seems to block the subject from any positive relation to the object judged, Hadot's cosmic consciousness comes up with a form of relation in feeling. The subject here is no longer the aesthete who judges the world disinterestedly, but the artist who attempts to put something new into it. The artist, Hadot says, "must paint in a state in which he *feels* his unity with the earth and with the universe." When the artist paints a work, he must explicitly identify himself with the same forces of nature that create the world and everything in it. This explicit identification is constituted in the feeling of the artist. Hadot here is relying on a number of writers on modern art who stipulate that art is itself the prolongation of the act of nature, and so is united with nature rather than separated from it. When Jackson Pollock, for example, was asked by a reporter why he painted abstract forms and did not take inspiration from nature, he replied: "I am nature." Lee Krasner, his wife and an artist in her own right, interpreted his response for the reporter: "he means he's total. He's undivided. He's one with nature, instead of 'that's nature over there and I'm here.'"[71] The center of Hadot's interest, however, seems not to lie in the prolongation of the act of nature, which could happen without any awareness on the part of the subject and so could not accomplish Hadot's goal of reuniting us with the world around us. Instead, Hadot focuses on the *feeling* of unity with nature brought about by the production of modern art.

Ecstasy

Hadot's interest in the artist's feeling of unity with nature borrows heavily from one of his favorite modern authors: Friedrich Nietzsche, and particularly the Nietzsche of the *Birth of Tragedy* (1872). Hadot refers to Nietzsche alongside Socrates and the Stoics in his own writings, and he also prepared an introduction to the French edition of Ernst Bertram's *Nietzsche: Versuch einer Mythologie.*[72] What could Nietzsche have offered Hadot? His *Birth of Tragedy* offers a portrait of emotion as the means to union with all things. Nietzsche famously calls this emotion "Dionysian," after the Greek god of wine, because the emotion that brings about unity with all things has its most basic form in intoxication.[73] It awakens "either under the influence of the narcotic draught, of which the songs of all primitive men and peoples speak, or with the potent coming of spring that penetrates all nature with joy."[74] When we get drunk or are overcome with emotion at the changing of winter into spring, "everything subjective vanishes into complete self-forgetfulness." Here lies the reason why emotion can bring about a unity of all things. It breaks down the barriers that the self erects against the outside world. When I get drunk, I no longer care to preserve my character, which distinguishes me from other human beings. I may even forget what that character was. Modes of interaction that preserve our distinction from one another, like conversation, and all the trappings of human convention, give way to unitary modes of action such as song and dance. The singers cannot be united only by the unity of the song they sing. They must lose themselves in a unity with the song, so that they no longer feel themselves separate from each other. In such a state, not only human beings feel themselves as one, but also "nature which has become alienated, hostile, or subjugated, celebrates once more her reconciliation with her lost son, man." The subjugation of nature to human purposes requires a self, capable of purposeful action, which distinguishes itself from nature as user from tool. But the Dionysian emotion has destroyed the self, and so nature too joins in this "mysterious primordial unity."

In its pure form, the Dionysian emotion uses no medium in the production of its art. The reveler himself "has become a work of art: in these paroxysms of intoxication the artistic power of all nature reveals itself." The unity of all nature, brought about by the intoxication of emotion, is the only pure Dionysian work of art. If this emotion is to be harnessed to a more enduring form of art, it must be combined with a medium. The medium of paint and canvas is not in itself Dionysian. The artist produces an image with it, and an image reinforces a series of distinctions: between spectator and image, between image and artist, between image and original. Rather than the cause or effect of unifying emotion, the image seems to be a barrier to such emotion. Nietzsche himself prefers to discuss music and theater as the media for Dionysian unity, and to leave painting and sculpture to the devices of a different impulse.[75] But we need not remove painting from the realm of the Dionysian if, like Pierre Hadot and the modern artists he follows, we downplay the painting itself in favor of its Dionysian production and Dionysian reception. The Dionysian emotion of unity with nature allows the painting to come forth as a work of nature rather than human intention. Because the image then functions as an incarnation of unifying emotion rather than a representation of some object, the spectator need no longer judge it as representing that object well or poorly. That is, he no longer need judge it as art. Nietzsche puts this new attitude of the spectator in the terms of tragic drama. Instead of the traditional spectator, who "must always remain conscious that he was viewing a work of art and not an empirical reality," we now have a spectator who "does not at all allow the world of the drama to act on him aesthetically, but corporeally and empirically."[76] The medium, in this case the actions performed on the stage, does not prompt the spectator to respond cognitively, but affectively, "corporeally and empirically," sharing in nature's visible reshaping of itself on the stage.

Architecture is another such medium for the Dionysian, though Nietzsche mentions it only occasionally in *The Birth of Tragedy* and elsewhere in his oeuvre. It is Lindsay Jones' *Hermeneutics of Sacred Architecture* that sets architecture in the context of ecstatic experience and makes a distinction similar to Nietzsche's between the image character of architecture that calls for a traditional spectator and the transformative experience to which it leads, engaging the spectator "corporeally and empirically."[77] Jones puts his distinction in the form of a commentary on Hans-Georg Gadamer but, as we will see, he steers Gadamer sharply in the direction of affective experience. Gadamer discusses architecture in the context of decoration, which in architecture is manifest above all in the building's facade. Decoration has two purposes in architecture: "to draw the attention of the viewer to itself, to satisfy his taste, and then to redirect it away from itself to the greater whole of the context of life which it accompanies."[78] The decoration of the architecture must be attractive enough to prevent "a dead or monotonous effect," but it must not get in the way of the building's purpose, which has its place in the context of the larger community. To this end, the decoration must neither physically interfere with the function of the building—making the doors too large to open comfortably, for example—nor isolate the building from its surroundings, perhaps by cutting off the entrance from the street.

Jones reinterprets Gadamer's two purposes of decoration to mean the following: "in order to have any transformative effect, a work of architecture must not simply elicit the admiration of viewers by reaffirming their 'tastes,'...but...the work must also possess what he terms 'the requisite enlivening effect,' that is, a component of variation, originality, or novelty that forces those viewers to readjust their tastes."[79] Jones interprets Gadamer's first condition for decoration—that it draw the attention of the viewer—as the representation of the familiar. We are drawn to things that have familiar associations for us. This interpretation departs significantly from Gadamer, who says that what delights us in decoration are "forms of nature used in an ornament," a notion apparently derived from Kant.[80] These forms of nature do not need to be recognized as the forms of particular natural objects, since we seek only the delight of the eye in following a complex tracery of lines.[81] On the second condition of the decorative—that it must redirect our attention to "the greater whole of the context of life"—Jones reads Gadamer's text in a way that goes directly against its meaning. Jones tells us that the redirection of our attention on the part of the decoration must "have an enlivening effect." The difficulty here is that Gadamer describes the enlivening effect as what draws attention to the decoration, not what redirects our attention away from it. Jones has put the "enlivening" on the wrong side of the dichotomy. Gadamer's decoration must avoid "a dead or monotonous effect," and so attract the viewer to itself, and only then must it redirect the viewer's attention away from itself, *after* it has exercised its "enlivening effect." This is not to say that the redirection of our attention to the larger context does not have an enriching effect on us, but it is not an effect of decoration. Instead, the building as a whole accomplishes a "heightening of a context of life"[82] by accomplishing purposes that help, rather than inhibit, our living together in a community. The exterior of a public building, for example, may face a park where public speeches and demonstrations may be held, rather than a busy street, which would impede all public gathering. This "heightening of a context of life" pertains to how the building accomplishes its purpose, a purpose that is always given within the context of our embodied interactions with each other. It does not pertain to its decoration. Jones, on the other hand, treats the redirection not as motion away from the decoration, but as motion away from the familiar side of decoration.[83] Our gaze remains drawn to the building, but we have been disturbed by its now unfamiliar character. Jones follows up on Gadamer's reference to the work of architecture's "strange" presentation of its original purpose with a string of references to the surrealist doctrine of art as "making strange."[84] We come to the building with certain expectations, and it draws us in with its surface appeal of familiarity, but it then redirects our experience in unfamiliar ways by the strangeness it subsequently reveals. Both the familiarity of the structure and its hidden strangeness are necessary for the building to signify something to us: "without a component of reassuring familiarity, the game (or the conversation) of a ritual-architectural event never gets started; but without a component of innovation or strangeness, nothing of significance transpires in those events."[85] This surreal strangeness of the architecture takes us quite some way from Gadamer's "forms of nature" and Kant's floral patterns. The building has now become capable, in the context of the ritual-architectural event, of producing a transformative experience in us. Part of Jones' difficulty in appropriating Gadamer here lies in the fact that Gadamer is discussing architecture in general, not the specific case of sacred architecture. As a result, Gadamer has no reason to discuss the experience of the building as salvific or transformative. Jones must import the language of the sacred from outside, but even here, Jones relies on a highly specific understanding of the sacred as that which "makes strange."

The affective experience of strangeness, which Jones characterizes as the significance of the ritual-architectural event, takes us far afield from the Neoplatonic characterization of the sacred. As we have seen, the Neoplatonists take little interest in the affective experience of the worshipper, preferring to speak of an ontological transformation brought about simply by the performance of the rite and, especially among the Christians, an intellectual contemplation that stands outside the affective. The weakness of the affective approach to the sacred is that, in its emphasis on forms of sensation, it risks overlooking the role of the body in the sacred place. We can see this encapsulated in Jones' creative reading of Gadamer. Where Gadamer's building provides a "heightening of a context of life" by accommodating itself to our bodies, Jones' building provides it by stimulating us to unusual and strange experience. It may do this through our bodily interaction with the place or it may not. The visitor to Apollo's oracle at Didyma, for example, sees from a distance what appears to be the entrance to the shrine. When he gets closer, he sees that the entrance is immense, and the step that leads to it is too large to be ascended by a human being. The entrance's lack of accommodation to a human body may produce in the visitor a feeling of the superhuman character of the place. On the other hand, a merely sensual stimulus, such as a "narcotic draught," could produce the same feeling.[86] In this case, there may be an experience of the sacred, but it has nothing to do with the sacred place.

CONCLUSION

RETHINKING THE SACRED PLACE

Our survey of sacred place in early medieval Neoplatonism has revealed two threads running through the philosophy of the period that are keenly relevant to the present-day rediscovery of place and sacred place. The first concerns Vincent Scully's dematerialization thesis: that Neoplatonic principles underlie the use of architecture to replace the material world with an immaterial one. We have some evidence of such principles. The Neoplatonists do think of some bodies as more material and others as less material, and they do think of the less material bodies as having more in common by nature with the divine intellect. But the theory of the sacred place developed by the later Neoplatonists does not make the dematerialization of bodies or of the place itself the defining characteristic of the sacred place. As Iamblichus and Dionysius both say, the sacred place is necessary because most human beings are only capable of interacting with the material. A sacred place that was less material would be less able to accomplish its task of connecting such people with the divine. Instead of directly revealing immaterial characteristics in immaterial bodies, the sacred place reveals the immaterial symbolically, through the thoroughly material bodies appropriate for most human beings. Within the ritual and mythic context of the sacred place, the material body acquires a significance beyond its nature, and allows the worshipper to live the divine life in the visible realm, or leads the worshipper to the contemplation of the divine through the visible realm. The later Neoplatonists, then, both Hellenic and Christian, do not think of the sacred place as replacing a crude, material nature with a more perfect, immaterial realm.

Scully notes that the transformation of Greek into Roman architecture involves a gradual walling-off of the exterior world in favor of an interior space entirely surrounded by products of human craft. In *Architecture: the Natural and the Manmade*, he describes this walling-off as a product of the Roman instinct "to enclose, to keep nature out, to trust in the man-made environment as a total construction."[1] We see a similar shift in the transformation of Hellenic into Christian sacred places. Where Hellenic temples are oriented toward exterior landscape features, are framed by columns instead of walls, and usually locate the sacrificial altar in the open air, Christian temples locate the altar and the rites within an interior space walled off entirely from the world outside. We have already seen Scully's example of the Hagia Sophia, which closes off an interior space and dematerializes it with its circular dome. Scully finds an example from the Roman rites in the Pantheon, which, seen from outside, is little more than a masonry shell, but which opens into a vast interior space.

Henri Lefebvre has criticized the claim that the Pantheon emphasizes interior space, though the target of his response is not Scully, who wrote his work fifteen years after Lefebvre's, but an argument presented most fully in Sigfried Giedeon's posthumously published work *Architecture and the Phenomena of Transition* (1971).[2] Giedeon does not oppose the Roman emphasis on interior space to the Greek orientation toward exterior nature, as Scully does. Giedeon's Romans do not close out nature; they close out social relationships like those grounded by the temple in Heidegger's "The Origin of the Work of Art." The Romans no longer make the temple the center of a world. While Egyptian and Greek architectural forms "were conceived and realized in the context of their social relationships—and hence from *without*," the Roman Pantheon "illustrates a second conception, under which the *interior* space of the monument became paramount."[3] Lefebvre responds that Giedeon has gotten the relationship between Greece and Rome backward. Relying on Heidegger's treatment of the Greek temple in "The Origin of the Work of Art," Lefebvre explains that the Greek temple grounds exterior social relationships only because it "encloses a sacred and consecrated space, the space of a localized divinity and of a divinized place, and the political centre of the city." It thus encloses an interior space as part of its grounding of the social relationships of the city. The enclosed space of the Roman Pantheon, too, does not prevent it from a relation to the exterior. The Pantheon, "as an image of the world or *mundus*, is an opening to the light; the *imago mundi*, the interior hemisphere or dome, symbolizes this exterior."[4] In other words, the literal enclosure of the space within the Pantheon has no meaning without the exterior world symbolized by it. The difference between Greece and Rome, then, does not arise from Greek openness to exterior space and Roman organization of interior space. For both the Greeks and Romans, the temple grounds a world that extends beyond it. The Roman temple is simply more universal in its symbolism. Where the Greek temple incarnates a specific god, whose power is exercised through a specific locality and specific things, the Roman temple is the symbol of all places and all gods: it is Pan-theon.

Neoplatonic reflection on the sacred place follows a similar development. The symbols employed in Iamblichan Neoplatonism are specific to certain things, certain people, certain places, and certain gods. As a result, it is vastly important that the worshipper know which god is allotted to him, and the places and things that are the allotments of that god. The context of the allotment, however, is ritual and symbolic, and does not depend on the natural relations between people, places, and things. For instance, owls may live around Athens. The people of Athens may build a shrine to Athena there, and take Athena as their patron. For the Athenians and the owls to manifest fully their natural forms, they must be in their proper places. The Athenian must be in Athens, and the owl must be in its tree. Standing on the steps of Athena's temple, one can look out and see everything in its proper place. But the ritual context has little to do with this exhibition of natural form, since it sees their relationships from the prior perspective of the god who establishes them. The Athenian, the owl, and the shrine of Athena are all together in the first place because they are peculiarly able to manifest Athena. While a temple that is open to the landscape is able visibly to relate these various symbolic things, people, and places to each other, the symbolic context remains even in a temple that is totally enclosed. In a Neoplatonic context, the walling-in of a sacred place need not imply the rejection of the exterior world. In the Christian Neoplatonist Maximus the Confessor, the walling-in of the sacred place may actually enhance its symbolic significance. Like most Christian Neoplatonists, Maximus has abandoned the allotment of specific temples to specific places with specific patrons. His church, like the Roman Pantheon as described by Lefebvre, is symbolic of the whole cosmos. Such a cosmos can never be seen in a landscape, since the landscape is always particular, and the cosmos is the totality of all things. The closing-off of the particular landscape can help to prevent the mistaking of the part for the whole, the particular landscape for the sum of all things, the world of a particular people for the universe that stands behind all peoples.

Which brings us to the second conclusion we may draw about the Neoplatonic sacred place. The symbolic relation of the sacred place to the universe does not necessitate that the world grounded by the sacred place extend as far as the universe. The Christian Neoplatonists do not initially consider the sacred place as a center on which to ground a Christian world. Dionysius seems to consider the altar as such a place, but the world that it grounds extends explicitly only as far as the walls of the church, and Dionysius says little about human life outside them. Alexander Golitzin has taken this exterior to be "an existence entirely sunk in deception and enslavement to the 'seeming good' of the world, the flesh, and the devil."[5] He bases this claim on the unhappy character of the catechumens, penitents, and possessed, who presumably inhabit the space outside the church. But the Christian laity and clergy also inhabit the space outside the church, and Dionysius never uses the negative language of Golitzin to describe it. All that seems to be excluded from this space is the explicitly symbolic, and therefore sacred, character of the rites. With Maximus the Confessor the Christian church, like the Roman Pantheon, becomes the symbol of the entire cosmos. But Maximus does not suggest that the context in which the church becomes symbolic should extend as far as the cosmos. His statements in the *Ambigua* on the division of nature suggest the opposite. In the cosmos as a whole, we human beings divide what ought to be united, and immerse ourselves in a part rather than uniting the parts to the whole. Even though by nature we can participate in a cosmic liturgy, our sin has made us incapable of it. We can only enter into the church—a part which is symbolic of the whole—and overcome the divisions of nature there.

This immersion in the part instead of the whole, and consequent treatment of the part as though it were the whole, becomes in Nicholas of Cusa the tragedy of religious and ethnic violence. A particular people in a particular place receives from its "kings and seers" a set of customs, which allow it to understand itself, God, and everything around it. This set of customs helps to shape the world of the historical people. And this people soon makes the mistake of taking the custom (the part) for the truth (the whole). The custom "is thought to have crossed over into nature." With this mistaken belief in hand, the people goes to war with other peoples, thinking of them as corrupt races worthy of destruction, or it attempts to convert them, thinking of them as capable of conversion to its truth. Cusa's solution is to apply a form of learned ignorance, whereby each people comes to realize that the truth of its rite is present in the rites of other peoples, since all rites are undergirded by a single religion. No single rite can lay claim to the one religion, since the religion is universal, and all rites are particular. Human beings should cultivate the presence of the divine in their own rite, and respect the religion that grounds all other rites, even though as a universal it cannot be seen in them. It can only be believed.

Cusa's account of the problems in the world-grounding aspect of the sacred place has a direct bearing on Martin Heidegger and Mircea Eliade's attempts to rediscover the sacred place in the twentieth century. As we have seen, both Heidegger and Eliade treat the sacred place as the visible ground of a world of meanings specific to a particular historical people. The fate of their own historical peoples illustrates the dangers of such an approach to the world more immediately for us than does the sack of Constantinople in 1453, which was so illuminating for Cusa. The alternative seems at first to be the restriction of the world grounded by the sacred place to the space within its walls. The resulting sacred place would look like what Eliade identifies as "a church in a modern city," which he describes without relish: "for a believer, the church stands in a different space from the street in which it stands. The door that opens on the interior of the church actually signifies a solution of continuity."[6] The exterior space of the church is now radically different from its interior space. Within is a context in which the divine itself becomes present. Without is the absence of meaning and divinity: a space devoid of the sacred. In the absence of any tie to the divine, the things outside the sacred place could easily degenerate into objects of merely instrumental value. Exterior space could then reasonably be organized for efficient use, promoting the homogeneous, geometrically ordered space against which Heidegger and Eliade react in the first place.

Heidegger himself discovers a *via media* between the consecrated world of the sacred place and the world of use objects, though it is really a rediscovery of the late medieval approach to things developed by, among others, Meister Eckhart. Where there is no consecrated place, things may still gather the divine into themselves, but they do so by making space for a god who remains absent, and manifests himself in the hidden nature of these things, or in their intrinsic value. The divinity encountered here is "the quiet desert into which distinction never gazed, not the Father, nor the Son, nor the Holy Spirit." Inside the church's walls, this divinity makes itself present as Father, Son, and Holy Spirit in the liturgy. Some division of exterior from interior space is essential to this *via media*, so that the space outside the church may avoid becoming the world of ethnic nationalism, and so that the space within may continue to be the place where God reveals his face.

ABBREVIATIONS

Amb.	*Ambiguorum Liber*
BT	*Being and Time*
BW	*Basic Writings*
Cat.	*Categories*
CCCM	*Corpus Christianorum Continuatio Medievalis*
CCSL	*Corpus Christianorum Series Latina*
CH	*On the Heavenly Hierarchy*
Conf.	*Confessions*
DDI	*On Learned Ignorance*
DDP	*Treatise on Divine Predestination*
De An.	*De Anima*
De Princ.	*De Principiis*
DM	*On the Mysteries of the Egyptians, Chaldeans, and Assyrians*
DN	*On the Divine Names*
EH	*On the Ecclesiastical Hierarchy*
Enn.	*Enneads*
Exp.	*Expositions on the Heavenly Hierarchy*
Gnost.	*Theological Chapters*
Hom. in Eccles.	*Homilies on* Ecclesiastes
In Cat.	*Commentary on Aristotle's* Categories
In Ioh. Ev.	*Commentary on the Gospel of John*
In Metaph.	*Commentary on Aristotle's* Metaphysics
In Prm.	*Commentary on Plato's* Parmenides
In Tim.	*Commentary on Plato's* Timaeus
Ioh.	Gospel of John
Is.	Book of Isaiah
Matth.	Gospel of Matthew
MT	*Mystical Theology*
Myst.	*Mystagogy*
Nic. Eth.	*Nicomachean Ethics*

PG	*Patrologia Graeca*
Phaed.	*Phaedrus*
Phys.	*Physics*
PL	*Patrologia Latina*
PLT	*Poetry, Language, Thought*
PP	*Periphyseon*
Prm.	*Parmenides*
Rep.	*Republic*
Thal.	*Questions for Thalassius*
Tim.	*Timaeus*

NOTES

Introduction

1. For a study of the Lysi frescos, and the story of how they made their way to Houston, see A.W. Carr and L.J. Morrocco, *A Byzantine Masterpiece Recovered, the Thirteenth-Century Murals of Lysi, Cyprus* (Austin: University of Texas Press, 1991). The Byzantine Fresco Chapel Museum, where the frescos now have their home, is the work of architect Francois de Menil.
2. On the chapel lighting, see K. Brady, "Houses of Worship," in *Lighting Design + Application*, vol. 29, no. 1 (1999), pp. 34–37.
3. *The Earth, the Temple, and the Gods: Greek Sacred Architecture* (New Haven and London: Yale University Press, 1962), p. 3. Scully's conclusions have not been universally accepted, and should be read in tandem with H.A. Thompson's critical review of his work in *Art Bulletin* 45, no. 3 (1963), pp. 277–80.
4. See, for example, Scully's analysis of the temple of Fortuna Primigenia at Praeneste: *Architecture: the Natural and the Manmade* (New York: St. Martin's Press, 1991), p. 109; *The Earth, the Temple, and the Gods*, pp. 210–12. See also Sigfried Giedeon's complementary account of the temple at Praeneste in *Architecture and the Phenomena of Transition: the Three Space Conceptions in Architecture* (Cambridge, Massachusetts: Harvard University Press, 1971), pp. 99–103.
5. *Architecture: the Natural and the Manmade*, p. 110.
6. *Architecture: the Natural and the Manmade*, p. 110.
7. *Architecture: the Natural and the Manmade*, p. 121.
8. *Architecture: the Natural and the Manmade*, p. 99. *The Earth, the Temple, and the Gods*, p. 187 does not refer to Neoplatonism by name, but only to Plato and Aristotle and their respective "mysticism and rationalism."
9. *The Crisis of European Sciences and Transcendental Philosophy: an Introduction to Phenomenological Philosophy*, tr. D. Carr (Evanston: Northwestern University Press, 1970), pp. 48–49.
10. *Parmenides*, tr. A. Schuwer and R. Rojcewicz (Bloomington and Indianapolis: Indiana University Press, 1992), p. 1.
11. *Parmenides*, p. 76.
12. *Parmenides*, p. 77.
13. On the ahistoricity of Heideggerian history, see J.D. Caputo, "Demythologizing Heidegger: *Aletheia* and the History of Being," in *Review of*

Metaphysics 41 (1988), pp. 519–46; P.W. Rosemann, "Heidegger's Transcendental History," in *Journal of the History of Philosophy*, vol. 40, no. 4 (2002), pp. 501–23.

14. Cited in R. Safranski, *Martin Heidegger: Between Good and Evil*, tr. E. Osers (Cambridge, Massachusetts: Harvard University Press, 1998), p. 249.
15. *The Sacred and the Profane: the Nature of Religion*, tr. W.R. Trask (New York: Harcourt, Brace and World, 1959), p. 162.
16. *The Sacred and the Profane*, p. 51.
17. *The Sacred and the Profane*, p. 50.
18. *The Sacred and the Profane*, p. 22. E. Durkheim discusses homogeneous space briefly in *The Elementary Forms of Religious Life*, tr. K.E. Fields (New York: The Free Press, 1995), pp. 10–11, from *Les formes élémentaires de la vie religieuse* (Paris: F. Alcan, 1912), but he does not develop a concept of sacred place in opposition to it, as Eliade does. Durkheim claims instead that *all* spatial representation, not just the religious variety, is "a primary coordination of given sense experience," and is culturally specific.
19. *The Sacred and the Profane*, p. 24.
20. *The Sacred and the Profane*, pp. 22–23.
21. *The Fate of Place* (Berkeley: University of California Press, 1997).
22. L. Jones, *The Hermeneutics of Sacred Architecture: Experience, Interpretation, Comparison* (Cambridge, Massachusetts: Harvard University Center for World Religions, 2000), vol. 1, p. xxviii. On the multiplicity of meanings in a building, see H. Lefebvre, *The Production of Space*, tr. D. Nicholson-Smith (Oxford: Blackwell, 1991), pp. 142, 160.
23. Dionysius did have predecessors who wrote about the Christian rites. P. Rorem, *Pseudo-Dionysius: a Commentary on the Texts and an Introduction to Their Influence* (New York, Oxford: Oxford University Press, 1993), pp. 118–21 provides a concise summary of their difference from Dionysius' own approach.
24. See "Apophatic-Kataphatic Tensions in Religious Thought from the Third to the Sixth Century A.D.: A Background for Augustine and Eriugena," in *From Augustine to Eriugena: Essays on Neoplatonism and Christianity in Honor of John O'Meara* (Washington, D.C.: Catholic University of America Press, 1991), p. 13, no. 3.
25. See *Oxford English Dictionary*, ed. J.A. Simpson and E.S.C. Weiner (Oxford: Clarendon Press, 1989), *s.v.* "pagan."
26. *The Fate of Place*, pp. 106–15.
27. Without a serious historical survey of the rise and fall of sacred place as a concept, we can expect a continuation of the trend of isolating figures from the Western tradition as ecological role models like L. White Jr.'s adoption of St. Francis as the "ecological saint" in "The Historical Roots of Our Ecologic Crisis," in *Science*, vol. 155, no. 3767 (March, 1967), pp. 1203–07 and A. Keselopoulos' adoption of St. Symeon the New Theologian for the same purpose in *Man and the Environment: a Study of St. Symeon the New Theologian*, tr. E. Theokritoff (New York: St. Vladimir's Seminary Press, 2001).

The Rediscovery of Place

1. On Heidegger's support for the National Socialists, see H. Ott, *Martin Heidegger: a Political Life*, tr. A. Blunden (New York: Basic Books, 1993); P. Adler, "A Chronological Bibliography of Heidegger and the Political," in *The Graduate Faculty Philosophy Journal*, vol. 14, no. 2 and vol. 15, no. 1 (1991), pp. 581–611; V. Farías, *Heidegger and Nazism*, tr. P. Burrell and G.R. Ricci (Philadelphia: Temple University Press, 1989). Eliade's support for the Iron Guard is recounted in A. Laignel-Lavastine, *Cioran, Eliade, Ionesco: L'Oubli du Fascisme* (Presses Universitaires de France, 2002); N. Manea, "Happy guilt: Mircea Eliade, fascism, and the unhappy fate of Romania," tr. A. Bley-Vroman, in *New Republic*, vol. 205, 6 (August 5, 1991), pp. 27–36.
2. *The Sacred and the Profane*, p. 23.
3. *The Fate of Place* (Berkeley: University of California Press, 1997), p. 284. For examinations of Heidegger's theory of place and its related terms, see *The Fate of Place*, pp. 243–84; D. Franck, *Heidegger et le Problème de l'Espace* (Paris: Editions de Minuit, 1986); Maria Villela-Petit, "Heidegger's conception of space," in *Critical Heidegger*, ed. C. Macann (London: Routledge, 1996), pp. 134–57.
4. *Phys*. 208b10; Heidegger, *Plato's Sophist*, tr. R. Rocjewicz and A. Schuwer (Bloomington and Indianapolis: Indiana University Press, 1997), p. 73. Heidegger's discussion here is examined in Stuart Elden, "The Place of Geometry: Heidegger's Mathematical Excursus on Aristotle," in *Heythrop Journal*, vol. 42, no. 3 (July, 2001), pp. 311–28.
5. *Phys*. 212a2.
6. *Plato's* Sophist, p. 74.
7. *Plato's* Sophist, p. 71.
8. *Plato's* Sophist, p. 75.
9. *Being and Time*, tr. J. Stambaugh (Albany: State University of New York Press, 1996), p. 102. Page numbers refer to the German edition, now published by Max Niemeyer (Tübingen, 1993).
10. *Plato's* Sophist, p. 72.
11. *Plato's* Sophist, p. 72. Aristotle, *Cat*. 6b2.
12. *BT* p. 103.
13. See "Conversation on a Country Path About Thinking," in *Discourse on Thinking* (New York: Harper and Row, 1966), pp. 58–90.
14. *Plato's* Sophist, p. 69.
15. *BT* p. 112.
16. *BT* p. 111.
17. "Building Dwelling Thinking," tr. A. Hofstadter, in *Basic Writings*, ed. D.F. Krell (New York: HarperCollins, 1993), p. 349.
18. *BW* p. 351.
19. See J.D. Caputo, "Demythologizing Heidegger: *Aletheia* and the History of Being," in *Review of Metaphysics*, 41 (1988), pp. 519–46; P.W. Rosemann,

"Heidegger's Transcendental History," in *Journal of the History of Philosophy*, vol. 40, no. 4 (2002), pp. 501–23.

20. *Rep.* 614c.
21. *Nic. Eth.* 1141b7; Heidegger, *Parmenides*, p. 100.
22. *Nic. Eth.* 1095a22–23.
23. *Parmenides*, p. 100.
24. *Parmenides*, pp. 100–101.
25. *Parmenides*, p. 101.
26. *Parmenides*, p. 102.
27. *Parmenides*, p. 102.
28. *Parmenides*, p. 102.
29. *Parmenides*, p. 102.
30. *Parmenides*, p. 103.
31. *Parmenides*, p. 104.
32. *Parmenides*, p. 109.
33. *Parmenides*, p. 107.
34. *Parmenides*, p. 117.
35. *Parmenides*, pp. 111–112.
36. In *Poetry, Language, Thought*, tr. A. Hofstadter (New York: Harper and Row, 1971), p. 132.
37. "The Origin of the Work of Art," tr. A. Hofstadter, in *BW*, p. 167.
38. *Parmenides*, p. 115.
39. See Heidegger's 1943 essay "Remembrance of the Poet," tr. W. Brock in *Existence and Being* (Chicago: Henry Regnery, 1949), p. 268, in which Heidegger describes kinship with the poets who name the gods as the "future of the historical being of the German people." See also *Elucidations of Hölderlin's Poetry*, tr. K. Hoeller (New York: Humanity Books, 2000), p. 171. On Heidegger's use of the poet Hölderlin in developing his concept of place, see S. Elden, "Heidegger's Hölderlin and the Importance of Place," in *Journal of the British Society for Phenomenology*, vol. 30, no. 3 (1999), pp. 258–74.
40. *BW*, p. 352.
41. *PLT*, p. 172.
42. *PLT*, p. 172. Heidegger also discusses consecration in the earlier "Origin of the Work of Art," in *BW*, p. 169.
43. *PLT*, p. 173.
44. *PLT*, p. 172.
45. *PLT*, p. 173.
46. *BW*, p. 354.
47. Heidegger seems to be referring to the latter secularity when he says that secularity is the degradation of the sacred in *Die Kunst und der Raum* (St. Gallen: Erker Verlag, 1969), tr. C.H. Siebert as "Art and Space" in *Man and World*, vol. 6, no. 1 (1973), pp. 3–5.
48. *BW*, p. 167.
49. *BW*, p. 168.

50. "*Religio in Stagno:* Nature, Divinity, and the Christianization of the Countryside in Late Antique Italy," in *Journal of Early Christian Studies*, vol. 9, no. 3 (2001), pp. 387–402. The above translations of Cassiodorus are also Barnish's.
51. On the basis of this transformation from Hellenic to Christian shrine, Barnish is able to argue against Peter Brown, who says in *The Cult of the Saints* (Chicago: University of Chicago Press, 1981), p. 126 that "the rise of Christianity in Western Europe is a chapter in the 'hominization' of the natural world." Brown employs in his work a different example of the Christian response to Hellenic rites: the sacrifices made on the edge of a marsh formed in a volcanic crater at Auvergne. When Gregory of Tours, bishop of Javols observed the rites, he advised that veneration instead be paid to the relics of Saint Hilary, to be found in a nearby church. Brown concludes from this that, though the pilgrimage to the crater continued, "the site itself is incorporated into an administrative structure dependent on the authority of human beings resident in a town far removed from the significant folds of the once holy landscape." Barnish points to the pond at Leucothea as a potent counter example.
52. See Trombly, *Hellenic Religion and Christianization* (Leiden: E.J. Brill, 1993), vol. 1, pp. 151–57; S.J.B. Barnish, "*Religio in Stagno:* Nature, Divinity, and the Christianization of the Countryside in Late Antique Italy," p. 391, n. 20. R. MacMullen, *Christianity and Paganism in the Fourth to Eighth Centuries* (New Haven and London: Yale University Press, 1997), pp. 64–65 and notes, provides a survey of Late Antique sources on the proscription of sacred trees, hilltops, and bodies of water; pp. 124–25 survey the transfer of such places from Hellenic to Christian hands.
53. "*Religio in Stagno,*" p. 392.
54. In *The Sacred and the Profane*, pp. 36–47; *Patterns in Comparative Religion* (New York: World Publishing, 1963), pp. 367–85.
55. In *Contesting the Sacred: the Anthropology of Christian Pilgrimage*, ed. J. Eade and M.J. Sallnow (Urbana and Chicago: University of Illinois Press, 2000), p. 9. See also L. Jones, *The Hermeneutics of Sacred Architecture* (Cambridge, Massachusetts: Harvard University Press, 2000), vol. 2, pp. 33–35.
56. *The Sacred and the Profane*, p. 27.
57. *Greek Religion*, tr. J. Raffan (Cambridge, Massachusetts: Harvard University Press, 1985), p. 113.
58. Plato, *Tim.*, 21c mentions Neith as the goddess who presides over the city of Saïs.
59. On places that are sacred for multiple peoples, see S. Coleman and J. Elsner, *Pilgrimage: Past and Present in the World Religions* (Cambridge, Massachusetts: Harvard University Press, 1995), pp. 48–51.
60. *The Sacred and the Profane*, p. 57.
61. D. Cave, *Mircea Eliade's Vision for a New Humanism* (New York: Oxford University Press, 1993), p. 19 notes that Eliade got this idea from Wolfgang

von Goethe's description of a "primordial plant," which orders all of the presently existing plants: "the primordial plant itself is an archetypal, metaphorical, symbolic concept having no historical actuality, except as it is manifested through the wide diversity of its variants in and throughout history." Cave later (p. 38) notes that this "is an aspect of symbols that is not appreciated enough in Eliade's thought. Interpreters instead focus on Eliade's emphasis on the transspatial and transtemporal quality of symbols."

62. *The Sacred and the Profane*, p. 137.
63. L. Jones, a self-described critical student of Eliade (see *The Hermeneutics of Sacred Architecture*, vol. 1, xxvi), catalogs numerous more recent attempts to find a common representational effect of sacred places, such as "homecoming (or homesickness), reunion, nostalgia, or *mal du pays*" (*The Hermeneutics of Sacred Architecture*, vol. 1, p. 76).
64. *Dramas, Fields, and Metaphors* (Ithaca and London: Cornell University Press, 1974), p. 189.
65. *Dramas, Fields, and Metaphors*, p. 226.
66. *Dramas, Fields, and Metaphors*, p. 182.
67. *Dramas, Fields, and Metaphors*, p. 201.
68. *Dramas, Fields, and Metaphors*, p. 227.
69. *Dramas, Fields, and Metaphors*, p. 201.
70. *Contesting the Sacred*, p. 5.
71. Hannah Arendt discusses the Greek *polis* as such a political place in *The Human Condition* (Chicago: University of Chicago Press, 1958), esp. pp. 196–99.
72. *Pilgrimage: Past and Present in the World Religions*, p. 202.
73. *Contesting the Sacred*, p. xiv.
74. *BT*, p. 102.
75. See *BT*, p. 105: "with the 'radio,' for example, Da-sein is bringing about today de-distancing of the 'world.'"
76. "The Nature of Language," in *On the Way to Language*, tr. P.D. Hertz (New York: Harper and Row, 1971), p. 103.
77. *On the Way to Language*, p. 105.
78. *On the Way to Language*, p. 103.
79. "Heidegger In and Out of Place," in *Heidegger: a Centenary Appraisal* (Pittsburgh: Silverman Phenomenology Center, 1990), pp. 62–98; *Getting Back into Place* (Bloomington and Indianapolis: Indiana University Press, 1993), pp. 130–32, and nn. 45 and 47 on p. 356; *The Fate of Place*, pp. 243–45. In *Being and Time*, p. 108 Heidegger says only that the corporeality of the human "contains a problematic of its own not to be discussed here."
80. *Getting Back into Place*, p. 45. Casey summarizes the history of place in Western philosophy in "Smooth Spaces and Rough-Edged Places: The Hidden History of Place," in *The Review of Metaphysics*, 51 (December, 1997), pp. 267–96.

81. *Phys.* 208b16–18. For a broader survey, see G.E.R. Lloyd, "Right and Left in Greek Philosophy," in *Journal of Hellenic Studies*, 82 (1962), pp. 56–66.
82. *Getting Back into Place*, p. 104.
83. Casey pays tribute to these four in *The Fate of Place*, p. 242.
84. *Getting Back into Place*, p. 46.
85. *Getting Back into Place*, p. 25.
86. *Getting Back into Place*, p. 101.
87. "The Relevance of the Beautiful," tr. N. Walker in *The Relevance of the Beautiful and Other Essays* (Cambridge: Cambridge University Press, 1986), p. 45.
88. *The Hermeneutics of Sacred Architecture*, vol. 1, p. 33.
89. *The Hermeneutics of Sacred Architecture*, vol. 1, p. 130.
90. *The Hermeneutics of Sacred Architecture*, vol. 1, p. 127. Jones notes that Catherine Bell, *Ritual Theory, Ritual Practice* (New York and Oxford: Oxford University Press, 1992), p. 44 suggests that even "the specific occurrence of a ritual" is not profitably treated as a text.
91. *From Text to Action*, tr. K. Blamey and J.B. Thompson (Evanston, Illinois: Northwestern University Press, 1991), p. 146. Cited by Jones in *The Hermeneutics of Sacred Architecture*, vol. 1, p. 126.
92. *The Production of Space*, p. 162.
93. *The Production of Space*, p. 162.
94. Several recent works have developed the social dimension of place and its shaping by power, including E.W. Soja's *Postmodern Geographies: the Reassertion of Space in Critical Social Theory* (Verso Books, 1997) and T.J. Gorringe's *Theology of the Built Environment* (Cambridge and New York: Cambridge University Press, 2002).
95. C. Long, *Significations* (Philadelphia: Fortress Press, 1986), p. 2 explains how the symbolic world can constitute "an arena and field of power relationships." See also J.Z. Smith's definition of religion and D. Cave's defense of Eliade from it in *Mircea Eliade's Vision for a New Humanism*, pp. 142–49.
96. *The Production of Space*, p. 135.
97. *The Production of Space*, p. 197.
98. *The Production of Space*, p. 198.

The Neoplatonic Background

1. R. Sorabji, *Matter, Space, and Motion: Theories in Antiquity and Their Sequel* (Ithaca, New York: Cornell University Press, 1988), mentions the Plotinian theory of place only on p. 206. S. Sambursky's *The Concept of Place in Later Neoplatonism* (Jerusalem: Israel Academy of Sciences and Humanities, 1982), admittedly a work on post-Plotinian Neoplatonism, gives only a few lines to Plotinus (pp. 15 and 39). Both works take their cue from P. Duhem's seminal account of the Neoplatonic view of place in

Le Système du Monde: Histoire des Doctrines Cosmologiques de Platon à Copernic (Paris: Hermann, 1913–1959), vol. 1, ch. 5.

2. *Prm.* 138a–b. See also *Tim.* 52c.
3. *Enn.* V, 5, 9, ll. 3–5.
4. Aristotle, on the other hand, claims in his *Physics* (210a22–24) that the being of an effect in its cause is different from, though analogous to, being in place.
5. *Enn.* V, 5, 9, ll. 16–18.
6. *The Fate of Place*, pp. 79–102.
7. *Phys.* 212b9–10.
8. On place in Proclus, see L. Schrenk, "Proclus on space and light," in *Ancient Philosophy* 9 (1989), pp. 87–94; L. Siorvanes, *Proclus: Neo-Platonic Philosophy and Science* (New Haven and London: Yale University Press, 1996), pp. 133–36, 247–56.
9. *The Concept of Place*, p. 66, ll. 1–25.
10. *The Concept of Place*, p. 66, ll. 27–28.
11. *The Concept of Place*, p. 66, ll. 29–31. Simplicius will be critical of this point among others. Light, according to Plato's *Timaeus*, is a species of the genus fire, and the species cannot be superior to the genus. Light, then, cannot be superior to fire. See *The Concept of Place*, p. 75, ll. 10–12. On Proclus' view of light and Simplicius' criticism of it here, see Siorvanes, *Proclus*, pp. 241–44.
12. *The Concept of Place*, p. 70, ll. 4–7.
13. *Life of Plotinus*, 13, ll. 6–7.
14. *DM*, III, 14.
15. *The Concept of Place*, p. 64, ll. 20–21.
16. We know that Proclus is not speaking about the place of the cosmos because of the claim that immediately follows: "the extension of the whole universe, the cosmic extension, differs from this particular extension."
17. *The Concept of Place*, p. 54, l. 25–26. The distinction is borrowed from Aristotle: *Phys.* 209a30–32.
18. *The Concept of Place*, p. 54, l. 30-p. 57, l. 1.
19. Contrast Aristotle, *Phys.* 209b22, who says that the proper place of a thing must be separable from it, and cannot be identified with its form.
20. *The Concept of Place*, p. 56, ll. 12–16. I follow Sambursky here in reading "in relation to it" (ὡς πρὸς ἐκεῖνο) instead of "just like it" (ὥσπερ ἐκεῖνο), the text found in *Simplicii in Aristotelis Physicorum Libros Quattuor Priores Commentaria*, ed. H. Diels (Berlin: George Reimer, 1882).
21. *The Concept of Place*, p. 56, n. 4.
22. *The Concept of Place*, p. 56, n. 5.
23. *The Concept of Place*, p. 53.
24. See *The Concept of Place*, p. 58–60, and especially p. 60, ll. 9–17 for a description of place as active extension in Syrianus.
25. *The Concept of Place*, p. 53.

26. This is the kind of place to which Aristotle's category of "whereness" (πoῦ) applies. See *Cat.* 2a1. I have borrowed the example of Socrates from the Neoplatonic commentaries on Aristotle's *Categories*, where it appears several times. Cf. Porphyry, *In Cat.* p. 77, ll. 22–23.
27. *The Concept of Place*, p. 42, ll. 1–2. Iamblichus here seems to rely on the famous claim of Pseudo-Archytas, that nothing can exist without place. See *The Concept of Place*, p. 36, ll. 3–4.
28. *The Concept of Place*, p. 42, ll. 6–10.
29. Simplicius takes time out from his summary of Iamblichus to notice this. See *The Concept of Place*, p. 42, ll. 15–17.
30. *The Concept of Place*, p. 42, ll. 19–22.
31. Is. 11:12. cf. Ioh. 11:52.
32. *The Concept of Place*, p. 16, n. 28. The phrase could just as easily have come from Heraclitus, fr. 91. Dionysius will use the phrase at *DN* 148, 9 (700A).
33. *The Concept of Place*, p. 48, ll. 3–7.
34. *The Concept of Place*, p. 48, ll. 29–31.
35. Sambursky adds "intelligible" to "divine place," a qualifier which is not in the Greek and which renders the phrase meaningless. God as place is higher than intellect.
36. *The Concept of Place*, p. 84, ll. 25–27.
37. *The Concept of Place*, p. 84, ll. 16–17.
38. Theophrastus' conception of place appears in Simplicius at *Phys.* 639, 15–22. See *The Concept of Place*, p. 32.
39. Damascius departs here from Aristotle, who denies that anything can properly be described as "in itself." See *Phys.* 210a25–b23.
40. *The Concept of Place*, p. 86, ll. 14–15.
41. *The Concept of Place*, p. 86, ll. 15–17.
42. Porphyry, *In Cat.* p. 77, ll. 25–27, calls the whole's being in the parts and the parts' being in the whole a different kind of "being in" than being in place. In this, he consciously or unconsciously follows Theophrastus, who also denies the relevance of the part/whole relation to place. Like Aristotle, Theophrastus prefers to restrict the name "place" to the inner surface of a being's surrounding body. See R. Sorabji, *Matter, Space, and Motion*, pp. 202–204.
43. *The Concept of Place*, p. 86, ll. 30–31.
44. *The Concept of Place*, p. 88, ll. 3–4.
45. *Enn.* V, 9, 11, l. 7.
46. *The Concept of Place*, p. 50, ll. 1–4.
47. *Rep.* 427b, tr. G.M.A. Grube, as revised by C.D.C. Reeve in *Plato: Complete Works*, ed. J.M. Cooper (Indianapolis and Cambridge: Hackett Publishing, 1997).
48. My summary here relies on my more thorough treatment in *The Problem of Paradigmatic Causality and Knowledge in Dionysius the Areopagite and His First Commentator* (Ann Arbor, Michigan: University Microfilms, 2001),

pp. 26–43. Proclus depends here on his teacher Syrianus, whose similar treatment begins at *In Metaph.* 107.5.

49. *In Prm.* p. 735, ll. 7–13. Page and line numbers refer to the edition of V. Cousin (Paris: Aug. Durand, 1864).
50. *In Prm.* p. 826, l. 13.
51. *In Prm.* p. 826.
52. For his earlier view, see *Enn.* II, 6, 1.
53. On the dating of the oracles, see H.D. Saffrey, "Les Néoplatoniciens et les oracles chaldaiques," in *Revue des Études Augustiniennes*, 27 (1981), pp. 209–25.
54. In *Porphyry's Place in the Neoplatonic Tradition: a Study in Post-Plotinian Neoplatonism* (The Hague: Martinus Nijhoff, 1974), p. 107, n. 11. cf. G. Shaw *Theurgy and the Soul: the Neoplatonism of Iamblichus* (University Park, Pennsylvania: Pennsylvania State University Press, 1995), pp. 162–66.
55. *DM* III, 15 (p. 136, ll. 7–8). Page numbers refer to the edition of G. Parthey (Berlin: F. Nicolai, 1857), and can be found in the now standard edition of E. des Places (Paris: Les Belles Lettres, 1966).
56. Fr. 22B93.
57. The bird may die as a result of the sign, by plunging into the ground, for example, but the sign itself does not kill the bird. The earth does. Likewise, the sign may indicate the imminent death of Julius Caesar, but the sign itself does not kill him. Brutus does.
58. *Enn.* I, 6, 1, ll. 20–22.
59. *Phaed.* 248d1–4 (tr. A. Nehamas and P. Woodruff in *Plato: Complete Works*, pp. 506–56).
60. *Enn.* I, 3, 1, ll. 26–27.
61. *Rep.* 401d (tr. G.M.A. Grube reprinted in *Plato: Complete Works*, pp. 971–1223).
62. *Enn.* I, 2, 2, ll. 14–18, and I, 2, 3, ll. 10–11.
63. *Enn.* I, 6, 3, ll. 17–19.
64. *Enn.* I, 6, 3, ll. 19–23.
65. *Enn.* III, 5, 9, l. 9.
66. *Enn.* V, 3, 6, ll. 15–16. See also III, 6, 1, l. 23; III, 7, 11, l. 20; III, 7, 12, l. 10; III, 7, 6, l. 39; I, 3, 4, l. 17; VI, 4, 15, l. 19. On species resting quietly in the "great intellect" which is their genus, see VI, 2, 20, l. 27.
67. *Enn.* V, 3, 7, ll. 14–16; III, 2, 2, l. 17. Plotinus uses "stillness" in its former sense of self-effacement to describe the divine mind when it is in contact with the first principle at *Enn.* VI, 9, 9, l. 18, and possibly VI, 9, 11, l. 13, though it is not clear whether intellect or the soul is the subject here.
68. *Enn.* V, 3, 12, ll. 35–36.
69. *Enn.* V, 1, 2, l. 14.
70. *Enn.* III, 2, 2, l. 15.
71. *Enn.* V, 1, 2, ll. 17–19.
72. "L'homme antique et la nature," reprinted in *Études de Philosophie Ancienne* (Paris: Les Belles Lettres, 1998), p. 311.

73. Iamblichus, *On the Pythagorean Way of Life*, tr. J. Dillon and J. Hershbell (Atlanta, Georgia: Scholar's Press, 1991), p. 120. See also Porphyry, *De Abstinentia*, I, 36, 1.
74. *The Philosophical History*, ed. Polymnia Athanassiadi (Athens: Apamea Cultural Association, 1999), p. 303.
75. *The Philosophical History*, pp. 35, 49.
76. "Le génie du lieu dans la Grèce antique," reprinted in *Études de Philosophie Ancienne*, pp. 319–24.
77. *Études de Philosophie Ancienne*, p. 322.
78. Porphyry, *Life of Plotinus*, sect. 10, l. 35.
79. See Armstrong's note on the passage in his translation of the *Life of Plotinus* (Cambridge and London: Harvard University Press, 1966). See also M. Edwards, *Neoplatonic Saints: the Lives of Plotinus and Proclus by their Students* (Liverpool: Liverpool University Press, 2000), p. 21, n. 115; P. Hadot, *Plotin ou la simplicité du regard* (Paris: Gallimard, 1997), p. 65.
80. *Lives of the Philosophers and Sophists*, published as Philostratus and Eunapius, *Lives of the Sophists*, tr. W.C. Wright (Cambridge, Massachusetts: Harvard University Press, 1921), p. 432.
81. *Lives of the Philosophers and Sophists*, p. 435.
82. *Enn.* IV, 8, 1, ll. 1–7.
83. *Enn.* IV, 8, 8, ll. 2–4.
84. *Enn.* IV, 8, 8, ll. 5–6.
85. See, for example, Philo of Alexandria, *De somniis*, 232. The passage is quoted in P. Hadot, *Plotin, Traité 9* (Paris: Les Éditions du Cerf, 1994), p. 205.
86. *Enn.* VI, 9. The same image appears also in V, 1, 6.
87. On this question, see M. Atkinson, *Ennead V, 1. A Commentary with Translation* (Oxford: Oxford University Press, 1981), pp. 133–34; P.A. Meijer, *Plotinus on the Good or the One* (J.C. Gieben: Amsterdam, 1992), p. 282, n. 805.
88. *Enn.* IV, 8, 8, ll. 1–3.
89. See C. Steel, *The Changing Self: a Study on the Soul in Later Neoplatonism* (Academie voor Wetenschappen, Letteren en Schone Kunsten: Brussels, 1978), p. 33.
90. The commentary survives only in fragments, the present one being no. 87, now collected and edited by John M. Dillon, *Iamblichi Chalcidensis in Platonis Dialogos Commentariorum Fragmenta* (Leiden: E.J. Brill, 1973). In the last proposition of his *Elements of Theology* (no. 211), Proclus makes the additional argument that the undescended soul erroneously confuses time and eternity. If part of the soul has not descended from the intelligible, it will think unceasingly either by always thinking the same thing, or by constantly shifting from one thought to another. If it always thinks the same thing, then it will be the intellect itself and not a part of the soul at all—since we as souls do not always think the same thing. If it constantly shifts from one thought to another, then a temporal thought

(constantly shifting) will be the same as eternal thought (the unceasing thinking), and so time will mix with eternity in a way that destroys the nature of both. If this is the case, then our attempt to exercise the activity of the undescended soul by removing all time and matter from our thoughts does not result in the experience of eternal intellect, but temporal privation.

91. *DM* V, 18 (p. 223, l. 10).
92. *DM* V, 18 (p. 223, l. 16–19).
93. At *In Prm.* 1071, Proclus says: "we do possess, inasmuch as we rank as souls, images of the primal causes, and we participate in both the whole soul and the plane of intellect and the divine henad" (tr. G.R. Morrow and J.M. Dillon).
94. *DM* V, 20 (p. 228, ll. 3–4, 6–8). Repeated at V, 22 (p. 230, l. 18–p. 231, l. 2).
95. On the term "theurgy," used by both Hellenic and Christian Neoplatonists, see H. Lewy, *Chaldean Oracles and Theurgy*, ed. M. Tardieu (Paris: Etudes Augustiniennes, 1978), pp. 461–66.
96. *DM* V, 22 (p. 231, ll. 2–4).
97. Parmenides acts as intellect to the more discursive soul of his conversation partner Zeno. See, for example, *Proclus' Commentary on Plato's* Parmenides, tr. G.R. Morrow and J.M. Dillon (New Jersey: Princeton University Press, 1987), p. 78: "Zeno constructed his arguments by this sort of dialectic, which combines propositions and notes consequences and contradictions; but Parmenides directly perceived the unity of being by using intellect alone, i.e, that intelligible dialectic which has its authority in simple intuitions."
98. *DM* V, 18 (p. 224, l. 11–13).
99. *Theurgy and the Soul*, pp. 189–215.
100. See Shaw, *Theurgy and the Soul*, p. 197.
101. See Shaw, *Theurgy and the Soul*, p. 201.
102. Shaw, *Theurgy and the Soul*, p. 170 suggests that names are a form of intermediate symbol, between the bodies of the material rite and the mathematical objects of the immaterial rite. Iamblichus, however, seems to locate names solidly within the material rite.
103. *DM* V, 19 (p. 225, ll. 13–15).
104. The phrase that follows makes little sense as it stands: "I mean the divine soul and nature, but not a pericosmic and genesiurgic soul and nature." Taylor modifies the Greek to read: "I mean from a divine, and not only from a mundane and genesiurgic soul and nature." This makes more sense than the Greek as it presently stands, given Iamblichus' overall point here, but it requires an unlikely transformation of the Greek.
105. "The law of worship assigns like to like" (p. 227, ll. 16–18).
106. *DM* V, 5 (p. 206, ll. 11–13).
107. *Enn.* IV, 5, 8, ll. 1–7.

108. A. Smith, *Porphyry's Place*, pp. 126–27 claims that, with this introduction of a second form of sympathy, "the continuity of Neoplatonic ontological progression breaks down." He contrasts this double sympathy with sympathy in Plotinus, for whom "the sympathy in the world on which divinization is based is the same as any other manifestation of sympathy."
109. *DM* V, 7 (p. 207, l. 15–p. 208, l. 1).
110. *DM* V, 8 (p. 209, ll. 9–10). The distinction between "accompanying cause" and "preceding cause" originates in Plato: *Tim*. 46c–d.
111. A. Smith, *Porphyry's Place*, p. 90 suggests that there are two kinds of theurgy, a lower kind which is "restricted to the area of συμπάθεια, the material world of humans and daemones," and a higher kind which "involves the linking of man with his superiors, the gods, not through συμπάθεια, but through φιλία." In view of Iamblichus' discussion here, we ought rather to say that all material theurgy has *philia* as its proper cause, and *sympatheia* as an "accompanying cause."
112. *DM* V, 14 (p. 217, ll. 11–12).
113. On the allotment in Iamblichus, see also fragment no. 14 of Iamblichus' commentary on the *Timaeus*, in Dillon, *Iamblichi Chalcidensis in Platonis Dialogos Commentariorum Fragmenta*, pp. 118–119. The language of allotment has a source in Plato's *Critias*, 109b–c, and *Timaeus*, 24c–d. Proclus comments on the latter passage in his *Commentary on Plato's* Timaeus, vol. 1, pp. 160–65. Volume and page numbers refer to the edition of E. Diehl (Leipzig: Teubner, 1903–06).
114. *DM* V, 9 (p. 30, ll. 13–15). Proclus distinguishes the way temples and cities receive divine power in *In Tim*, vol. 1, p. 125.
115. *DM* I, 9 (p. 33, ll. 7–9).
116. *DM* V, 23.
117. *DM* V, 24 (p. 235, ll. 5–9).
118. *Histories*, II, 65–76.
119. *DM* I, 13 (p. 43, ll. 4–8).
120. *DM* III, 15 (p. 135, ll. 2–3). Plato compares the human arts of prophecy unfavorably with the ecstatic prophecy of the oracles in his *Phaedrus*, 244c–d.
121. *DM* III, 15 (p. 135, ll. 13–14).
122. Walter Burkert, *Greek Religion*, p. 112, has distinguished the Greeks from the Romans on this latter point. Both cultures of bird interpretation rely on a prophet sitting in a fixed seat that faces north, regarding all motion left as bad and right as good, but the Greeks rely more on the intuition of the particular prophet, while the Romans tend more to codify the principles of bird observation. Whether Burkert is correct on this point or not, the difference between the Greeks and Romans is posterior to the more elemental distinction between the image and the sign.
123. *DM* III, 16 (p. 136, ll. 16–19).
124. *DM* III, 16 (p. 137, ll. 6–10).

125. *DM* III, 25 (p. 158, ll. 5–6).
126. *The Earth, the Temple, and the Gods*, p. 131. "Hexastyle" means "having six columns in front." The "pronaos" is a columned porch outside the cella or main hall of the temple.
127. *The Earth, the Temple, and the Gods*, p. 130. Scully refers to the enclosed valley as a "natural megaron."
128. Burkert, *Greek Religion*, p. 115.
129. Thomas Taylor, the English translator of *On the Mysteries*, consistently mistranslates "priest" as "priestess." Usually the prophet was female, but at Claros it was male. See Tacitus, *Annals*, II, 54: "it is not a woman, as at Delphi, but a priest chosen from certain families, and almost always from Miletus who receives the visitors."
130. *DM* III, 11 (p. 125, ll. 15–17).
131. *The Earth, the Temple, and the Gods*, p. 131.
132. I mean by "magic" here the manipulation of occult sympathies between objects to produce desired effects in the visible world without involving the soul's salvation.
133. *Greek Religion*, p. 116.

Dionysius the Areopagite

1. *The Earth, the Temple, and the Gods: Greek Sacred Architecture* (New Haven and London: Yale University Press, 1962).
2. In some cases, new myths have appeared to maintain the sanctity of existing sites within a new language community. The oldest site at Claros, for example, is not associated with Apollo. See Scully, *The Earth, the Temple, and the Gods*, pp. 130–31.
3. *The Earth, the Temple, and the Gods*, p. 1.
4. *The Earth, the Temple, and the Gods*, p. 3.
5. Originally published in German as *Griechische Religion der archaischen und klassischen Epoche* (Stuttgart: W. Kohlhammer, 1977); English translation by J. Raffan (Cambridge, Massachusetts: Harvard University Press, 1985), p. 85.
6. *Greek Religion*, p. 86.
7. *Greek Religion*, p. 269.
8. *Greek Religion*, p. 270.
9. *The Earth, the Temple, and the Gods*, p. 2.
10. *The Earth, the Temple, and the Gods*, p. 46.
11. *Greek Religion*, p. 91.
12. *Early Christian and Byzantine Architecture*, 4th ed. (New York: Penguin, 1986), p. 95.
13. *The Earth, the Temple, and the Gods*, p. 3.
14. *De Architectura*, III, 8, 6.
15. (Manchester: Manchester University Press, 1968). Chapter four gives special attention to landscape features, and discusses several of the passages I provide later.

16. *Ep.* XIV.
17. *Enn.* vol. 5 (Cambridge: Harvard University Press, 1984), p. 14. Armstrong also cites two other references to the Plotinian cosmos visualization exercise in Basil, but the first of these (found in the homily *De Fide*, at *PG* 31, 465A12–B1) does not seem dependent on Plotinus, since it does not mention stillness, and does not privilege the four regions of the cosmos as its subject. The second citation refers to a passage that clearly does depend on Plotinus, but it is from the fifth book of *Adversus Eunomium* (*PG* 29, 769A13–B4), which is no longer considered to be the work of Basil.
18. *Ep.* IV, 6.
19. *Ep.* V, 5.
20. *The Greek Patristic View of Nature*, p. 90.
21. "Byzantine Monastic Horticulture: the Textual Evidence," in *Byzantine Garden Culture*, ed. A. Littlewood, H. Maguire, and J. Wolschke-Bulmahn (Washington, D.C.: Dumbarton Oaks, 2002), pp. 37–67.
22. *The Life and Miracles of St. Luke of Steiris*, tr. C.L. and W.R. Connor (Massachusetts: 1994), ch. 54. Cited in "Byzantine Monastic Horticulture," p. 38.
23. *Acta Sanctorum* (Paris: V. Palme, 1863–1940), November, vol. 3, 520E.
24. See James of Kokkinobaphos (mentioned on p. 107 of *Byzantine Garden Culture*), who describes the garden as "tranquil." H. Maguire notes in the same volume that later Byzantine gardens tended to follow the route of Christian temples. Their gardens were enclosed rather than open to the surrounding landscape (pp. 31, 35).
25. *Conf.* IX, 10.
26. See *Enn.* I, 3, 1, where Plotinus explains that making philosophical arguments is itself a path to the divine.
27. *Conf.* VIII, 2.
28. B. McGinn, "From Admirable Tabernacle to the House of God," in *Artistic Integration in Gothic Buildings* (Toronto, 1995), pp. 41–56 suggests that Augustine regards the visible church as a means of access to heaven, despite the fact that Augustine never says any such thing: "I do not think it out of harmony with Augustine's thought (though it goes beyond anything he said in explicit fashion) to claim that, from an Augustinian perspective, the church building is essentially a *tabernaculum admirabile* whose function is to give access to the *domus Dei* of heaven."
29. *Hom. I (De Incomprehensibilitati)*, 202–206, *PG* 48, 705B.
30. In John Chysostom, *Sur l'Incompréhensibilité de Dieu*, tr. R. Flacelière (Paris: Les Éditions du Cerf, 1970), p. 37. cf. also *The Greek Patristic View of Nature*, p. 94.
31. *Hom. in Eccles.* viii (*PG* 44, 729D–732A)
32. *DM* V, 15.
33. *EH* 87, 21–24 (433D–436A).
34. *Ep. IX*, 198, 8–9 (1108A).

35. Dionysius frequently refers to these examples from sense experience as "paradigms" (παραδείγματα), following a longstanding Platonic tradition. See M. Harrington, *The Problem of Paradigmatic Causality and Knowledge in Dionysius the Areopagite and His First Commentator* (Ann Arbor, Michigan: University Microfilms, 2001), pp. 19–21.
36. *DN* 195, 4–6 (868B).
37. *DN* 185, 16–18 (821B).
38. *DN* 153, 17–154, 1 (705B).
39. See Proclus, *In Prm.* Book I, *passim.*
40. *DN* 140, 4 (681B).
41. *DN* 159, 4–6 (712A). Dionysius here quotes Gal. 2:20.
42. *DN* 159, 6–8 (712A). Paul calls himself "effaced" at II Cor 5:13.
43. *DN* 133, 14–134, 1 (648A–B).
44. *DN* 134, 2–4 (648B).
45. *EH* 81, 10–13 (428A).
46. *EH* 73, 7–8 (397A).
47. *DN* 140, 17–20 (681C).
48. *DN* 140, 8–10 (681B).
49. *EH* 73, 15–16; 74, 2–3 (397B–C).
50. *EH* 74, 10 (397C).
51. As Dionysius does at *CH* 37, 9 (260B).
52. *EH* 126, 18–127, 1 (561A–B).
53. Iamblichus devotes the entirety of *On the Mysteries*, book nine, to the subject of the guardian angel (ἴδιος δαίμων). While he may possibly have been influenced by the Judaic tradition here, there is also a Platonic source of the doctrine.
54. *CH* 39, 5–8 (261B).
55. Gen. 14:18.
56. *CH* 38, 17–20 (261A).
57. *De Princ.* I, 8, 1 and III, 3, 3. Also *Contra Celsum* V, 30. See J. Danielou, *Les Anges et leur mission* (Chevetagne, 1952), pp. 25–36.
58. *EH* 67, 3–4 (376B).
59. *EH* 67, 11–12 (376C).
60. *EH* 84, 11–13 (432A–B).
61. *EH* 67, 6–7 (376B).
62. *EH* 79, 7–12 (424C).
63. The object of this contemplation is, as Dionysius says at *EH* 74, 10–11 (397C), the "principle and knowledge of what is sacral according to sense."
64. *EH* 130, 3–4 (565B).
65. *L'univers dionysien: structure hiérarchique du monde selon le Pseudo-Denys* (Paris: Les Éditions du Cerf, 1983), p. 259.
66. *L'univers dionysien,* p. 260. For a general treatment of architecture in Dionysius, see P. Scazzoso, *Ricerche sulla Struttura del Linguaggio dello*

Pseudo-Dionigi Areopagita (Milan: Società Editrice Vita e Pensiero, 1967), pp. 133–49; A.M. Schneider, "Liturgie und Kirchenbau in Syrien," in *Nachrichten der Akademie der Wissenschaft in Göttingen* (1949), pp. 56–68; C. Schneider, "Καταπέτασμα," in *Kyrios* 1 (1936), pp. 57–73.

67. *Early Christian and Byzantine Architecture* (New York: Penguin, 1986), p. 203.
68. *The Early Churches of Constantinople: Architecture and Liturgy* (University Park and London: Pennsylvania State University Press, 1971), pp. 117–34.
69. *EH* 71, 5–7 (393C).
70. *EH* 92, 10–12 (441D).
71. The hierarch is also described as "gathering the sacred choir" for the funeral rites: *EH* 122, 23 (556C).
72. Plotinus, too, treats the activities of light and fragrance as parallel. See *Enn.* V, 1, 6, ll. 28–38.
73. *EH* 98, 24–26 (477C).
74. *EH* 99, 1–3 (477C–D).
75. *EH* 85, 17 (433A).
76. *CH* 9, 1–2 (121D).
77. *EH* 82, 15–16 (428D–429A). The censing procession is also described at 476D.
78. *EH* 84, 1–6 (429D-432A).
79. Dionysius never speaks of a union brought about through taste, though the members of the hierarchy taste the bread and wine.
80. *EH* 88, 12 (437A).
81. *EH* 88, 19 (437A).
82. See Proclus, *In Prm.* 810, ll. 37–39 and 912, ll. 32–35. He is commenting on Plato's use of "like in like" at *Prm.* 132D.
83. *EH* 95, 9–12 (473A).
84. *EH* 124, 16–17 (557C). Similar statements can be found in the other passages dealing with expulsion of the catechumens, penitents, and possessed.
85. (Toronto: Pontifical Institute of Medieval Studies, 1984), p. 7, n. 17. See also p. 109.
86. *Pseudo-Dionysius: a Commentary on the Texts and an Introduction to Their Influence* (New York: Oxford University Press, 1993), p. 122.
87. *The Early Churches of Constantinople*, p. 144.
88. *EH* 76, 18–20 (401B).
89. *EH* 82, 15–17 (428D–429A).
90. *EH* 82, 18–20 (429A).
91. *EH* 97, 24 (476D).
92. *EH* 82, 22–83, 2 (429A).
93. *EH* 91, 13–14 (441B).
94. *EH* 91, 11–13 (441A–B).
95. *EH* 91, 22–92, 1 (441B–C).
96. *EH* 82, 5–6 (428C).

97. *EH* 81, 6–7 (425D); (440B); 92, 17 (444A).
98. *EH* 111, 10–15 (509D).
99. *EH* 114, 15–16 (516B).
100. *EH* 103, 4–5 (484D).
101. *Ep.* VIII, 176, 13–177, 1 (1088D).
102. *EH* 122, 23–123, 2 (556C).
103. *EH* 81, 11–13 (428A).
104. At *MT* 146, 13 (1033B). Place appears among the denials of sensible categories at *MT* 148, 3 (1040D).
105. *DN* 207, 10–208, 1 (909B).
106. On emanation in Dionysius and the Neoplatonic tradition, see the summary in S. Gersh, *From Iamblichus to Eriugena: an Investigation of the Prehistory and Evolution of the Pseudo-Dionysian Tradition* (Leiden: E.J. Brill, 1978), pp. 17–26.
107. Compare Augustine's argument at *Conf.* VII, 15.
108. *MT* 148, 3 (1040D).
109. *DN* 138, 10 (680B).
110. Ex. 24:10, Septuagint version.
111. A. Golitzin, *Et Introibo ad Altare Dei* (Thessalonica: George Dedousis, 1994), pp. 168–77 discusses the liturgical significance of Moses' ascent.
112. *MT* 143, 18–19 (1000C).
113. See my explanation of this passage in *A Thirteenth-Century Textbook of Mystical Theology at the University of Paris* (Leuven: Peeter's Press, 2004). The more traditional explanation is that the underlying structures are God himself. See V. Lossky, "La théologie négative dans la doctrine de Denys l'Aréopagite," in *Revue des sciences philosophiques et théologiques*, 28 (1939), p. 218; Y. de Andia, *L'union à Dieu chez Denys l'Areopagite* (Leiden: Brill, 1996), pp. 344–48.
114. *CH* 32, 7–8 (212C).
115. *DM*, III, 25 (p. 158, ll. 5–6).
116. *DM*, III, 25 (p. 159, ll. 17–19).
117. See A. Papaconstantinou, "Oracls chrétiens dans l'Egypte byzantine: Le témoignage des papyrus," *Zeitschrift für Papyrologie und Epigraphik* 104 (1994), pp. 281–86; R. MacMullen, *Christianity and Paganism in the Fourth to Eighth Centuries* (New Haven and London: Yale University Press, 1997), pp. 121–23, 138–39 with n. 127.
118. *DN* 120, 11–12 (597A).
119. He says that union with the divine darkness is "beyond intellect" at *MT* 144, 15 (1001A). He calls it an "ecstasy" at *MT* 142, 10 (1000A).
120. *DN* 133, 14–134, 4 (648A–B).
121. *EH* 92, 10–12 (441D).
122. *DN* 159, 9–11 (712A–B).
123. *DN* 159, 12–14 (712B).
124. *DN* 115, 16–18 (593A).
125. *De An.* 402b21–22.
126. *DN* 192, 3–5 (856D).

127. *Ep.* IV, 160, 12–161, 2 (1072B).
128. Matth. 14:25.

The Dionysian Tradition

1. *John of Scythopolis and the Dionysian Corpus: Annotating the Areopagite* (Oxford: Clarendon Press, 1998), pp. 11–22.
2. At *De hominis opificio*, sect. 16 (*PG* 44: 180A).
3. In dating Maximus' works, I follow Polycarp Sherwood's *An Annotated Date-List of the Works of Maximus the Confessor* (Rome: Orbis Catholicus, Herder, 1952).
4. *Amb.* 41 (*PG* 91: 1304D–1316A).
5. *Amb.* 41 (*PG* 91: 1305A).
6. *Amb.* 41 (*PG* 91: 1305B7). On Maximus' use of both literal and figurative language when speaking about paradise, see L. Thunberg, *Microcosm and Mediator: the Theological Anthropology of Maximus the Confessor* (Chicago: Open Court, 1995), pp. 381–91.
7. *Amb.* 10, sect. 19 (*PG* 91: 1136C11–12).
8. *Myst.* (*PG* 91: 668C–669A).
9. *Amb.* 41 (*PG* 91: 1312B).
10. *Gnost.* I. 68 (*PG* 90: 1108C). cf. II, 32. God transcends place at I, 69–70.
11. *Amb.* 41 (*PG* 91: 1305A–B).
12. *Amb.* 41 (*PG* 91: 1305C).
13. *Amb.* 41 (*PG* 91: 1305D).
14. *Amb.* 41 (*PG* 91: 1305D).
15. *Thal.* XXXI, ll. 9–12. Maximus also describes the soul as a church at *Thal.* LXIV, ll. 806–807 (*PG* 90: 728A).
16. *Thal.* XXXI, ll. 12–17.
17. *CH* 32, 7–8 (212C).
18. *Amb.* 10, sect. 38 (*PG* 91: 1180A–B).
19. *Amb.* 10, sect. 38 (*PG* 91: 1180C). See John of Damascus, *Exposition of the Christian Faith*, sect. 13.
20. *Amb.* 10, sect. 38 (*PG* 91: 1180C).
21. *Amb.* 41 (*PG* 91: 1308C).
22. For Maximus' treatment of the Platonic genera, see M. Harrington, "Creation and Natural Contemplation in Maximus the Confessor's *Ambiguum* X.19," in Omnia in Sapientia: *Essays in Honour of R.D. Crouse* (Leiden: E.J. Brill, forthcoming).
23. *Amb.* 41 (*PG* 91: 1313B).
24. *Amb.* 41 (*PG* 91: 1313A).
25. *Amb.* 41 (*PG* 91: 1312B).
26. Not for nothing did Hans Urs Von Balthasar put the single word ἀσυγχυτῶς, meaning "unconfused," at the head of *Cosmic Liturgy*, his now classic work on Maximus the Confessor.
27. On the comprehensive glance in Plotinus, see J. Phillips, "Plotinus and the 'Eye' of Intellect," in *Dionysius* XIV (1990), pp. 79–103.

28. *Myst.* (*PG* 91: 664D, 668C).
29. *Pauline and Other Studies in Early Church History* (London: Hodder and Stoughton, 1906), p. 131.
30. J.-C. Larchet, *La divinisation de l'homme selon saint Maxime le Confesseur* (Paris: Les Éditions du Cerf, 1996), p. 401, n. 15 cautions that when Maximus calls the church an image, "it evidently does not concern the church considered as a building, as one might believe by understanding the text literally. The architectural topography of the church symbolizes here the spiritual topography of the church."
31. *Myst.* (*PG* 91: 665C2–4).
32. *Amb.* 41 (*PG* 91: 1308C8–10).
33. *Myst.* (*PG* 91: 665C5–10).
34. *Myst.* (*PG* 91: 668A3–6).
35. *Myst.* (*PG* 91: 668C10).
36. *Myst.* (*PG* 91: 669A–B).
37. *Myst.* (*PG* 91: 672A3–4).
38. *Myst.* (*PG* 91: 672C).
39. *Myst.* (*PG* 91: 681D).
40. *Myst.* (*PG* 91: 669A).
41. *Myst.* (*PG* 91: 681D–684A).
42. *Annotated Date-List*, p. 27.
43. *Ep.* 22 (*PG* 91: 605).
44. *Amb.* 41 (*PG* 91: 1308A). On Maximus' use of the term "interval" or "extension," see Thunberg, *Microcosm and Mediator*, pp. 57–60.
45. Sherwood, *Annotated Date-list*, p. 32.
46. *Ep.* 24 (*PG* 91: 608–13).
47. *Ep.* 27 (*PG* 91: 617–20).
48. The Acts of the Apostles also uses the phrase "one soul" to describe the unity of the Christians at Pentecost. See Acts 4:32.
49. I Cor. 15:28.
50. The letter is addressed to "Conon, priest and superior."
51. *Ep.* 25 (*PG* 91: 613).
52. *Ep.* 8 (*PG* 91: 440–45).
53. See, for instance, Evagrius' *Praktikos*, chs 6–14 on "the eight provocations."
54. For general treatments of place in Eriugena, see: M. Cristiani, "Le problème du lieu et du temps dans le livre Ier du 'Periphyseon,' " in *The Mind of Eriugena*, ed. J.J. O'Meara and Ludwig Bieler (Dublin: Irish University Press, 1973), pp. 41–47; D. Moran, *The Philosophy of John Scottus Eriugena: a Study of Idealism in the Middle Ages* (Cambridge: Cambridge University Press, 1989), pp. 194–99.
55. *DDP* IX, 6, ll. 134–39. Line numbers refer to G. Madec, ed. *De Divina Praedestinatione Liber*, in *CCCM*, vol. 50 (Turnholt: Brepols, 1978).
56. The cliché can be traced back at least as far as Plotinus, *Ennead* VI. 7. 31, ll. 24–25, who speaks of things "scattered by magnitudes" (*μεγέθεσι διειλημμένα*).

57. *PP* II, p. 13, ll. 122–26 (*PL* 122: 533B). Page and line numbers refer to E. Jeauneau, ed. CCCM vols. 61–65 (Turnholt: Brepols, 1996–2003).
58. *PP* V, p. 126, ll. 4069–70 (*PL* 122: 950C).
59. *PP* V, p. 138, ll. 4481–86 (*PL* 122: 959A–B).
60. *PP* V, p. 41, ll. 1282–86 (*PL* 122: 888C).
61. *PP* V, p. 42, ll. 1308–10 (*PL* 122: 889A). Also quoted by Eriugena in his earlier *DDP*, VIII, 7, p. 53, ll. 148–50. In this earlier work, Eriugena does not regard Augustine's use of place as posing a problem for other authorities of the church. It is representative, then, of what we may call his "pre-critical" period.
62. *PP* V, p. 44, ll. 1355–59 (*PL* 122: 890A).
63. *PP* V (*PL* 122: 888B, 906A)
64. *PP* V, p. 43, ll. 1348–49 (*PL* 122: 889D).
65. *PP* II, p. 83, ll. 1972–77 (*PL* 122: 586D).
66. *PP* I, p. 47, ll. 1375–76 (*PL* 122: 474B).
67. *PP* I, p. 49, ll. 1432–33 (*PL* 122: 475C).
68. *PP* I, p. 52, ll. 1564–65 (*PL* 122: 479A).
69. *PP* I, p. 51, ll. 1533–36 (*PL* 122: 478B).
70. The *nutritor* says at *PP* I, p. 54, ll. 1625–27 (*PL* 122: 480B): "we see by *μετονομία*, that is, a transfer of name, that those things which are contained are named after those things that contain them."
71. See the rest of the liberal arts discussion at *PL* 122: 766C and 869A.
72. *PP* V, p. 44, ll. 1358–59 (*PL* 122: 890A).
73. I Cor. 15:44.
74. *PP* III (*PL* 122: 729C–730A). The two interlocutors sometimes say that the human soul is the image of God by being a trinity of mind, vital motion, and matter (*PP* IV, *PL* 122: 790B–C), a triad pioneered by Gregory of Nyssa. They call this trinity a "second image" of God, since the first image lies in the triad of the soul's motions and does not include the body at all. The degree to which we include the body in the image of God, of course, determines the degree to which we make the body essential to human existence after death.
75. This is already the case for the elements in their uncompounded state. See *PP* V (*PL* 122: 902A–B).
76. *PP* III, p. 159, ll. 4659–63 (*PL* 122: 730C).
77. The *nutritor* adds smell to the list of favored senses at *PL* 122: 731A.
78. *PP* V, p. 186, ll. 6046–48 (*PL* 122: 993C–D).
79. *PP* IV, p. 47, ll. 1267–69 (*PL* 122: 773B).
80. *PP* V, p. 51, ll. 1586–88 (*PL* 122: 895A).
81. *PP* V, p. 50, ll. 1572–73 (*PL* 122: 894C).
82. *PP* V, p. 60, ll. 1917–20 (*PL* 122: 901D–902A).
83. *PP* V, p. 28, ll. 830–32 (*PL* 122: 879A).
84. *PP* V, p. 177, ll. 5762–67 (*PL* 122: 986D).
85. Quoted by Eriugena at *PP* IV, p. 109, ll. 3275–79 (*PL* 122: 818C).

86. *PP* IV, p. 129, ll. 3940 (*PL* 122: 833A).
87. *PP* IV, p. 142, ll. 4324–27 (*PL* 122: 841B).
88. *PP* IV, p. 131, ll. 4001–04 (*PL* 122: 834B). See also 829C.
89. *PP* III, p. 91, ll. 2640–48 (*PL* 122: 683A–B).
90. *PP* I (*PL* 122: 454A–B).
91. *PP* III, p. 91, ll. 2648–50 (*PL* 122: 683B).
92. Ioh. 4: 5–42.
93. *In Ioh. Ev.*, p. 318, ll. 59–63 (*PL* 122: 339A). Augustine, *In Ioh. Ev.* tract. XV, 26, in *CCSL*, vol. 36, p. 161, ll. 14–18 (*PL* 35, 1519–20) interprets the passage to mean that we should ourselves become the temple of God, but he nowhere uses the term "intellect."
94. *In Ioh. Ev.*, p. 316, ll. 50–52 (*PL* 122: 338D).
95. *PP* V, p. 169, ll. 5517–22 (*PL* 122: 981B–C).
96. *Exp.* ed. J. Barbet, in *CCCM*, vol. 31 (Turnholt, Brepols, 1975), p. 16, ll. 571–75.
97. *Exp.* p. 17, ll. 581–82.
98. The major study of Eriugena's eucharistic theology remains R.U. Smith, "*Oratio Placabilis Deo:* Eriugena's Fragmentary Eucharistic Teaching in the light of the Doctrine of the *Periphyseon*," in *Dionysius* XIII (1989), pp. 85–114.
99. *PP* III, p. 92, ll. 2665–67 (*PL* 122: 683C).
100. II Cor. 12:2.
101. *De Administratione*, p. 62, ll. 29–30. Discussed by Panofsky on p. 21.
102. *De Administratione*, p. 62, l. 31–p. 64, l. 3.
103. On the metaphysics of light in Suger, see Panofsky's bibliography in *Abbot Suger: on the Abbey Church of St.-Denis and its Art Treasures* (Princeton, New Jersey: Princeton University Press, 1946), p. 165.
104. *De Administratione*, p. 50, ll. 10–11.
105. *De Administratione*, p. 46, l. 27–p. 48, l. 1.
106. G. Zinn, "Suger, Theology, and the Pseudo-Dionysian Tradition," in *Abbot Suger and Saint-Denis*, ed. P.L. Gerson (New York, 1986), pp. 33–40; P. Kidson, "Panofsky, Suger, and St. Denis," in *Journal of the Warburg and Courtauld Institutes* 50 (1987), pp. 1–17; B. McGinn, "From Admirable Tabernacle to House of God," in *Artistic Integration in Gothic Buildings* (Toronto, 1995), pp. 41–56.
107. "Suger, Theology, and the Pseudo-Dionysian Tradition," p. 35.
108. "Suger, Theology, and the Pseudo-Dionysian Tradition," p. 34.
109. "Suger, Theology, and the Pseudo-Dionysian Tradition," pp. 36–37.
110. *De Administratione*, p. 74, l. 2.
111. "From Admirable Tabernacle," pp. 48–49.
112. "From Admirable Tabernacle," p. 48.
113. *De Consecratione*, pp. 100–101, and McGinn, "From Admirable Tabernacle," p. 49 and n. 25. O. von Simson discusses the possible origins of Gothic architecture in the medieval theory of geometrical proportion in *The Gothic Cathedral: Origins of Gothic Architecture and the Medieval Concept of Order* (New York, 1956), pp. 21–50.

114. "From Admirable Tabernacle," pp. 49–50.
115. Plotinus' argument against suicide rests on this principle, that the violence done to the body is not justified by the increased activity of the intellect after death releases it from the body. See *Enn.* I, 9.
116. *DM* V, 12.
117. *DM* III, 14.
118. *Architecture: the Natural and the Manmade*, p. 123. Von Simson calls Gothic "transparent, diaphanous architecture" in *The Gothic Cathedral*, p. 4.
119. See *Enn.* I, 6, 1, l. 33, where gold accompanies the sun, lightning, and stars as examples of things whose beauty does not derive from proportion.

The Loss of Place

1. "Building Dwelling Thinking," in *BW*, p. 352.
2. L. Versényi, *Heidegger, Being, and Truth* (New Haven and London: Yale University Press, 1965); J. Caputo, *The Mystical Element in Heidegger's Thought* (Athens: Ohio University Press, 1978). More recent comparisons of Heidegger and Eckhart include S. Sikka, *Forms of Transcendence: Heidegger and Medieval Mystical Theology* (Albany: State University of New York Press, 1997), esp. pp. 143–86; V. Vitiello, " 'Abgeschiedenheit', 'Gelassenheit,' 'Angst': Tra Eckhart e Heidegger," in *Heidegger and Medieval Thought* (Turnhout: Brepols, 2001), pp. 305–16; G. Strummiello, " 'Got(t)heit': la Deità in Eckhart e Heidegger," in *Heidegger and Medieval Thought*, pp. 339–59.
3. *The Principle of Reason*, tr. R. Lilly (Bloomington and Indianapolis: Indiana University Press, 1991), p. 38.
4. *The Mystical Element in Heidegger's Thought*, p. 65.
5. *The Mystical Element in Heidegger's Thought*, p. 86.
6. German sermon #48, tr. E. Colledge in *Meister Eckhart: the Essential Sermons, Commentaries, Treatises, and Defense* (New Jersey: Paulist Press, 1981), pp. 197–98.
7. On Cusa's use of one-point perspective, see K. Harries, *Infinity and Perspective* (Cambridge, Massachusetts: MIT Press, 2001).
8. *DDI* III. 3 (h I, 195–202). Numbers in parenthesis refer to the volume and paragraph of Heidelberg Academy edition of Cusa's works: *Nicolai de Cusa Opera omnia iussu et auctoritate Academiae Litterarum Heidelbergensis* (Hamburg, 1932–). The English translation in *Nicholas of Cusa: Selected Spiritual Writings*, tr. H.L. Bond (New York and Mahwah: Paulist Press, 1997), preserves this paragraph numbering.
9. The term "recapitulation" is the apostle Paul's, used by him to describe the work of Christ at Eph. 1:10.
10. *On Conjectures*, II, 14 (h III, 144).
11. *On Conjectures*, II, 14 (h III, 143).
12. *DDI* III. 1 (h I, 189).
13. *On the Peace of Faith*, 1.

14. *On the Peace of Faith*, 2.
15. *On the Peace of Faith*, 2.
16. *On the Peace of Faith*, 4.
17. *On the Peace of Faith*, 6. James Biechler and Lawrence Bond, *On Interreligious Harmony* (Lewiston: Mellen, 1990), p. 222, n. 13 suggest that Cusa derived this idea from his reading of the Qur'an.
18. Cusa does not go so far as to approve of all existing rites without modification. Most of the discussion between the representatives of God and the representatives of the various rites involves coming to an agreement that many Christian doctrines ought to replace less perfect doctrines already present in the other rites. While Cusa accepts or rejects components of other rites according to whether they violate the Christian creed, we can imagine other criteria for evaluating rites, such as the one provided by D. Cave in *Mircea Eliade's Vision for a New Humanism* (Philadelphia: Fortress Press, 1993), p. 24: "symbols are validated or rejected by whether or not they can survive repeated valuations and devaluations in history and by whether they can integrate a plurality of symbols."
19. *On the Vision of God*, 1 (h VI, 5). cf. *De Deo Abscondito*, 14: *deus dicitur a theoro*.
20. *On the Vision of God*, 2 (h VI, 7).
21. *On the Vision of God*, 5 (h VI, 13).
22. The seeing of God is the place of God, which is paradise. Paradise is then an intelligible place. See *On the Vision of God*, 9 (h VI, 37).
23. *On the Vision of God*, 6 (h VI, 18).
24. Cusa could draw the three faces from either the three non-human faces of Ezekiel's vision, or from Xenophanes, fr. 21B15.
25. *DDI* II, 11 (h I, 156).
26. On God as place in Cusa, see C.L. Miller, "Meister Eckhart in Nicholas of Cusa's 1456 Sermon: *Ubi est qui natus est rex Iudeorum?*" and E. Brient, "Meister Eckhart and Nicholas of Cusa on the 'Where' of God," in *Nicholas of Cusa and His Age: Intellect and Spirituality*, ed. T.M. Izbicki and C.M. Bellitto (Leiden: Brill, 2002), pp. 105–25 and 127–50.
27. *DDI* II, 11 (h I, 161).
28. *DN* 146, 14–15 (697B).
29. *DDI* II, 12 (h I, 169–72).
30. Unlike Heidegger, Cusa does not hold that non-human animals are worldless.
31. On the limited sense in which Cusa acknowledges space to be infinite, see *DDI* II, 1 (h I, 97).
32. See J. O'Neill, "The Varieties of Intrinsic Value," in *The Monist* (1992), p. 131: "it is possible to talk in an objective sense of what constitutes the good of entities, without making any claims that these ought to be realized."
33. *Should Trees Have Standing?—Toward Legal Rights for Natural Objects* (Los Altos, California: William Kaufmann, 1974). In what follows I will mean by intrinsic value the second of the three kinds identified by J. O'Neill in "The Varieties of Intrinsic Value," pp. 119–20.

34. Holmes Rolston identifies the species, the ecosystem, the earth, and nature itself as candidates for such larger systems of value in "Value in Nature and the Nature of Value," in *Philosophy and the Natural Environment*, ed. R. Attfield and A. Belsey (Cambridge: Cambridge University Press, 1994), pp. 13–30.
35. "An Amalgamation of Wilderness Preservation Arguments," in *The Great New Wilderness Debate*, ed. J.B. Callicott and M.P. Nelson (Athens, Georgia: University of Georgia Press, 1998), pp. 154–98.
36. *The Great New Wilderness Debate*, p. 191.
37. *The Sacred and the Profane*, p. 24.
38. For the dependence of the unfolding on nature, see *On Conjectures*, II, 14 (h III, 143–44). For its dependence on circumstances, see *DDI* III, 1 (h I, 189).
39. *BW*, pp. 168–75.
40. *BW*, p. 171.
41. *BW*, p. 173.
42. *BW*, p. 172.
43. In the summer of 1942, Heidegger taught a lecture course on Hölderlin's poem entitled "The Ister." "Ister," Heidegger tells his audience, is the Roman name for the lower Donau river, and much of the course explores the essence of rivers. Part of the exploration involves a consideration of the river as "enigma" (*Rätsel*). See M. Heidegger, *Hölderlin's Hymn "The Ister,"* tr. W. McNeill and J. Davis (Bloomington and Indianapolis: Indiana University Press, 1996), from *Hölderlins Hymne>>Der Ister<<* (Frankfurt am Main: Vittorio Klostermann, 1984).
44. *Parmenides*, p. 128.
45. "The Question Concerning Technology," in *BW*, p. 321.
46. *BW*, p. 168.
47. *The Principle of Reason*, p. 42.
48. Heidegger speaks of the material of language in "The Origin of the Work of Art," in *BW*, pp. 171 and 173.
49. *Life of Plotinus*, 11 (p. 37, ll. 13–16).
50. *DM* III, 9 (p. 116, ll. 9–11).
51. *Life of Pythagoras*, 25, p. 136.
52. *DM* III, 9 (p. 118, ll. 13–15). E. des Places gives two commentaries on this chapter.
53. *DM* III, 9 (p. 118, l. 17–p. 119, l. 4).
54. *Parmenides*, p. 102.
55. *Parmenides*, p. 101.
56. *Parmenides*, p. 106.
57. *Parmenides*, p. 106.
58. *The Hermeneutics of Sacred Architecture*, vol. 1, p. 95. Jones here refers to Gadamer's *Truth and Method* (New York: The Seabury Press, 1975), p. 100, where Gadamer is not talking about architecture.
59. *Truth and Method*, p. 124.

60. *Truth and Method*, p. 124. I have corrected the English version, which makes no sense as it stands. The German, p. 133 reads: "durch die Darstellung erfährt es gleichsam einen *Zuwachs an Sein*."
61. *The Hermeneutics of Sacred Architecture*, pp. 95–97.
62. Hadot's concept of the charming, as the object of a disinterested judgment, maps onto Kant's concept of the beautiful, described in Book One of the *Critique of Judgment*. The concept of the sublime is covered in Book Two of the same work. See esp. sect. 23, where Kant explicitly distinguishes the beautiful from the sublime.
63. See *Critique of Judgment*, sects. 2–4.
64. "L'homme antique et la nature," reprinted in *Études de philosophie ancienne*, p. 307.
65. Heidegger criticizes the aesthetic approach to architecture at *Parmenides*, p. 115. Gadamer does the same at *Truth and Method*, p. 139.
66. *Letters to Lucilius*, 64, 6. Translated in P. Hadot, *Philosophy as a Way of Life*, tr. M. Chase (Oxford and Cambridge, Massachusetts: Blackwell, 1995), p. 257. Hadot also discusses it in *Études de philosophie ancienne*, p. 314.
67. See, for example, Epictetus' *Discourses*, I, 4, 1; III, 12, 8; III, 13, 21; III, 22, 13; IV, 4, 33; *Handbook*, 2; *Fragments*, 27.
68. *Philosophy as a Way of Life*, p. 252.
69. *Philosophy as a Way of Life*, p. 252.
70. *Philosophy as a Way of Life*, p. 255.
71. See S. Naifeh and G.W. Smith, *Jackson Pollock: an American Saga* (New York: C.N. Potter, 1989), p. 486, referring to an article by Amei Wallach in *Newsday*, Sept. 23, 1981.
72. *Nietzsche: Essai de mythologie* (Paris: Le Félin, 1990), pp. 5–44.
73. In the first few paragraphs of the *Birth of Tragedy*, Nietzsche says that he calls the emotion "Dionysian" because music is its primary artistic stimulus, and Dionysius is the god of music. But Nietzsche immediately turns to the more fundamental pre-artistic experience of the Dionysian: intoxication. It turns out that music is only an artistic way of bringing about the emotion whose more original form is drunkenness, presided over by Dionysius as god of wine. Dionysius becomes the god of music because he is the god of wine.
74. *The Birth of Tragedy Out of the Spirit of Music*, tr. W. Kaufmann in *Basic Writings of Nietzsche* (New York: Modern Library, 1992), p. 36.
75. Namely, the Apollinian. See *The Birth of Tragedy*, p. 33.
76. *Birth of Tragedy*, p. 57.
77. Vincent Scully also finds an experiential component to architecture while discussing one of the most recent movements in North American architecture, the New Urbanism. He notes in Peter Katz' *The New Urbanism: Toward an Architecture of Community* (New York: McGraw Hill, 1994), p. 225 that the New Urbanism aims to satisfy both of the ways we experience works of visual art: "empathetically and by association. We feel them both in our bodies and in terms of whatever our culture has taught us."

Modern architecture often pursues geometrical forms to break the second form of experience, by removing the building from traditional styles which move us because they enfold us within our own culture. While the community of Seaside, Florida, perhaps the most famous example of the New Urbanism, does not follow modern architecture to the extreme of mere geometrical forms, and in its overall pattern embraces a traditional method of design, Scully leaves us with a sense that it is the empathetic experience of architecture that has the most hold on him at Seaside: "when the great winds rise up out of the Gulf—and the storm clouds roll in thundering upon the little lighted town with its towered houses—then a truth is felt, involving the majesty of nature and, however partial, the brotherhood of mankind" (*The New Urbanism*, p. 230). We sense that the sundering of the unity between natural landscape and human artifice after the period of the great Greek temples has the capacity to be partially healed at Seaside, but the means by which this healing occurs now has a distinctly emotional character.

78. *Truth and Method*, p. 140.
79. *The Hermeneutics of Sacred Architecture*, p. 63.
80. See Kant's description of ornaments at *Critique of Judgment*, sect. 14, in tandem with his discussion of free beauty in sect. 16.
81. *Truth and Method*, p. 140. If we critically adopt a Kantian position here, and decide that our delight in complex forms is derived from a cultural habit, rather than the nature of our minds, then the distinction between Gadamer and Jones here may not seem so great.
82. *Truth and Method*, p. 139.
83. Strangely, Jones cites here and quotes in his endnotes the work of J. Weinsheimer, who correctly and impressively interprets Gadamer here.
84. Gadamer, *Truth and Method*, p. 138 says that the building must preserve its original "contexts of purpose and of life" if it is to be a work of art. It cannot simply abandon its purpose or the lives of the people who will inhabit it. Its artistic character lies only in making these purposes seem "strange" in the completed work. Jones, *The Hermeneutics of Sacred Architecture*, p. 64 refers to the surrealist doctrine of "making strange," and he refers to the surrealists themselves on p. 66.
85. *The Hermeneutics of Sacred Architecture*, p. 71.
86. Nietzsche, *The Birth of Tragedy*, p. 36.

Conclusion

1. *Architecture: the Natural and the Manmade*, p. 111.
2. *Architecture and the Phenomena of Transition: the Three Space Conceptions in Architecture* (Cambridge, Massachusetts: Harvard University Press, 1971). Lefebvre's notes refer to Giedeon's *Space, Time, and Architecture*

(Cambridge: Harvard University Press, 1967) here, but this work mentions the Pantheon only once and in a different context.

3. *The Production of Space*, II. 8, p. 127.
4. In fairness to Giedeon, we should note that he says much the same thing in *Architecture and the Phenomena of Transition*, p. 148: "this 'templum deorum omnium' stands as a symbol of the all-embracing power, the cosmic extent, of the Roman Empire."
5. *Et Introibo ad Altare Dei*, p. 158.
6. *The Sacred and the Profane*, p. 25.

INDEX